BSA Singles Owners Workshop Manual

by Marcus Daniels

Models covered:
The BSA unit construction singles
First introduced in 1958
Discontinued in 1972
247 cc C15, B25, C25
343 cc B40
441 cc B44
499 cc B50

ISBN 978 0 85696 127 4

Printed in Malaysia   *(127-9AP10)*

ABCDE
FGHIJ
KLMNO

3

Haynes Publishing Group
Sparkford Nr Yeovil
Somerset BA22 7JJ England

Haynes Publications, Inc
859 Lawrence Drive
Newbury Park
California 91320 USA

Acknowledgements

Our thanks are due to BSA Motor Cycles Limited for their assistance. Brian Horsfall gave the necessary assistance with the overhaul and devised many ingenious methods for overcoming the lack of service tools. Les Brazier arranged and took the photographs that accompany the text. Tim Parker advised about the way in which the text should be presented.

Our thanks are also due to Mike Lane, of Mike Lane Motor-cycles, Downton, Nr Salisbury, who supplied the B25 "Starfire" featured in the photographs accompanying the text, and to Paul Wright who allowed us to photograph his Victor Special for the front cover. Finally we would also like to acknowledge the help of the Avon Rubber Company, who kindly supplied illustrations and advice about tyre fitting, and Amal Limited for their carburettor illustrations.

About this manual

This manual has been compiled to enable the owner of a BSA "single" to carry out normal routine maintenance and if necessary major repair work. It has been assumed that the average owner does not own a highly sophisticated workshop. Therefore, unless absolutely essential, BSA Service Tools are not prescribed and an alternative method using tools found in the average household are used.

Each of the six chapters is divided into numbered sections. Within the sections are numbered paragraphs. Cross-reference throughout this manual is quite straightforward and logical. When reference is made, 'See Section 6.10' - it means section 6 paragraph 10 in the same Chapter. If another Chapter were meant it would say 'See Chapter 2, Section 6.10'.

All photographs are captioned with a section/paragraph number to which they refer, and are always relevant to the chapter text adjacent.

Figure numbers (usually line illustrations) appear in numerical order, within a given chapter. 'Fig. 1' therefore, refers to the first figure in Chapter 1.

Left hand and right hand descriptions of the machines and their components refer to the left and right of a given machine when normally seated facing the front wheel.

Motorcycle manufacturers continually make changes to specifications and recommendations, and these, when notified, are incorporated into our manuals at the earliest opportunity. Whilst every care is taken to ensure that the information in this manual is correct no liability can be accepted by the authors or publishers for loss, damage or injury, caused by any errors in or omissions from the information given.

Differences in the BSA "Single" range

The range of BSA single cylinder, unit construction machines are very similar in design, ie the 250 cc and 350 cc Star models have much in common, as do the 441 cc B44 and the later 500 cc B50 models. However, where there is a significant difference between models the separate dismantling/assembly procedure is described under the appropiate heading. .

Although this manual does not specifically cover the 250/Triumph Trophy models, owners of these machines will find this book extremely informative as the engines of the 250 Triumph and BSA 'Starfire' are virtually identical and the stripdown/rebuilding photographs will provide a useful guide for the home maintenance owners of 'Trophy' motorcycles.

Contents

BSA Gold Star 500 SS

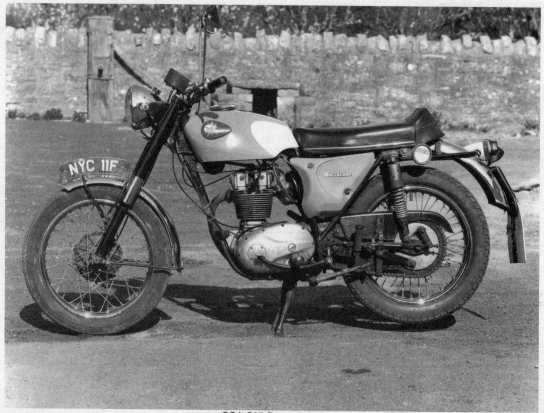

BSA C25 Barracuda

BSA C15 250

BSA B25 Starfire 250

Introduction to the BSA OHV 'single'

In 1910 the Birmingham Small Arms Company changed from bicycle, to motorcycle manufacture with the introduction of two belt driven 500 cc single cylinder machines: the standard model developing 3½ horsepower and the tuned version 4½ horsepower. The engines however were sidevalve and it was not until 1923 that the first OHV model was introduced - a vertical single of 350 cc. In 1927 BSA brought out a single cylinder 500 cc OHV 'sloper' model featuring a duplex frame, internal expanding brakes and a three speed gearbox.

By 1930 a vertical 350 cc single and a vertical and 'sloper' 500 cc single were being offered to the public. A tuned version of the 500 cc 'sloper' was also available - on the timing case of which was a red star. Producing 24 horsepower at 5250 rpm, this machine was capable of over 80 mph and was the first of a long and successful line of BSA Star models. Towards the end of 1930 the 'slopers' were discontinued and 350 cc and 500 cc vertical models were produced; tuned versions of these machines being named the Blue Star 1935 saw the introduction of the 250, 350 and 500 cc OHV Empire Star models and on June 30th 1937 a 500 cc version of this machine was entered at Brooklands and won its race at an average speed of over 100 mph - the award for this event was the Gold Star.

When a new 500 cc single was developed the following year, it was of course named the Gold Star; producing 30 hp at 5800 rpm it was capable of speeds up to 110 mph when stripped for racing - and, because of the low price and high performance large numbers were sold. In 1939 the Empire Star series were replaced by the 250 cc C10 and 350 and 500 Silver Star models.

With the event of the second world war motorcycle production at the BSA factory came to a halt, 1949 saw the re-introduction of the Gold Star model with a plunger frame. By 1954 this model had been developed into a very sophisticated road/racing machine comprising an all alloy engine embodying large cooling fins, fitted into a double downtube frame with swinging arm rear suspension. It continued a highly successful career until 1963 when production ceased - much to the disappointment of many thousands of Gold Star enthusiasts all over the world.

1958 saw the beginning of a complete departure from the usual engine/transmission design with the introduction of unit construction machines in place of the 250 cc C12, the 350 B31 and the 500 B33. The first of these new breed of motorcycles, the 250 cc C15 Star, was an instant success - it had a clean, uncluttered appearance and provided reliable and economical transport. The 350 cc B40 followed in 1960 and in 1965 BSA offered the powerful 441 cc B44 to be superseded by the 500 cc B50, which, in a moment of nostalgia was named the Gold Star.

During 1967 the 250 cc C15 was replaced by the B25 Starfire and C25 Barracuda models, which featured a quickly detachable rear wheel and 12 volt lighting.

Trials versions of virtually all the BSA single cylinder range of motorcycles have been produced, one of the most outstanding being the B44, which became very popular with some of the top riders, due to its excellent power/weight ratio and handling qualities.

These latter years of motorcycle production have seen the innovation of the three and four cylinder "superbikes" which develop more power than most small saloon cars. In the 100 to 250 cc range, two-stroke and overhead cam engine design has reached the stage where 5000 rpm is often stall speed and the red line on the tachometer is set at a figure more in keeping with the revolutions of a jet turbine. Despite this, the simple ruggedness of a "one lunger" will always have an immense appeal for those riders who consider the high torque and reliability of a 'single' to be unbeatable.

DIMENSIONS

Model	B25SS	B25T	B44	B50SS	B50T	B50MX
Wheelbase (in.)	54.0	54.0	53.0	54.0	54.0	54.0
Ground clearance (in.)	7.0	7.5	8.0	7.0	7.5	7.5
Seat height (in.)	32.0	32.0	32.0	32.0	32.0	32.0
Overall width (in.)	29.0	29.0	27.0	29.0	29.0	33.5
Overall length (in.)	85.0	85.0	83.0	85.0	85.0	82.5
Overall height (in.)	43.5	43.5	42.0	43.5	43.5	43.5
Dry weight (lbs)	290.0	287.0	320.0	310.0	298.0	240.0

Ordering spare parts

When ordering spare parts for any of the BSA single cylinder range, it is advisable to deal direct with an official BSA agent, who will be able to supply many of the items ex-stock, even though a particular machine may be out of production. Parts cannot be obtained direct from BSA Motor Cycle Limited - all orders must be routed through an approved agent, even if the parts required are not held in stock.

Always quote the engine and frame numbers in full, particularly if the parts are required for any of the earlier models. Include any letters before or after the number itself. The frame number will be found stamped on the left hand side of the front engine mounting lug and the engine number is on the left hand side of the crankcase, immediately below the cylinder barrel. On some models the frame number is located either on the prop-stand lug or on the left side of the steering head.

Use only parts of genuine BSA manufacture. Pattern parts are available but in many instances they will have an adverse effect on performance and/or reliability. Some complete units are available on a 'service exchange' basis from certain large specialist dealers, (the factory system no longer functions); particularly in London. The usual parts covered are components such as the barrel and crankshaft - it is certainly an economic method of repair. Retain any broken or worn parts until a new replacement has been obtained. Often these parts are required as a pattern for identification purposes, a problem that becomes more acute when a machine is classified as absolute. In an extreme case, it may be possible to reclaim the broken or worn part, or to use it as the pattern for making a new replacement. Many older machines are kept on the road in this way, long after a manufacturer's spares have ceased to be available.

Some of the more expendable parts such as spark plugs, bulbs, tyres, oils and greases etc., can be obtained from accessory shops and motor factors, who have convenient opening hours, charge lower prices and can often be found not far from home. It is also possible to obtain parts on a Mail Order basis from a number of specialists who advertise regularly in the motor cycle magazines.

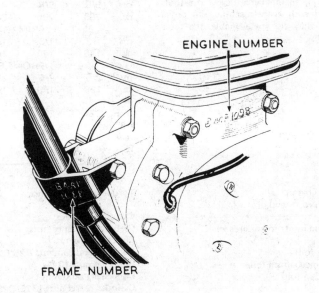

Frame and Engine numbers location

Routine maintenance

Periodic routine maintenance is a continous process that commences immediately the machine is used. It must be carried out at specified mileage recordings or on a calendar basis if the machine is not used frequently, whichever falls soonest. Maintenance should be regarded as an insurance policy, to help keep the machine in the peak of condition and to ensure long, trouble-free service. It has the additional benefit of giving early warning of any faults that may develop and will act as a regular safety check, to the obvious advantage of both rider and machine alike.

The various maintenance tasks are described under their respective mileage and calendar headings. Accompanying diagrams are provided, where necessary. It should be remembered that the interval between the various maintenance tasks serves only as a guide. As the machine gets older or is used under particularly adverse conditions, it would be advisable to reduce the period between each check.

Some of the tasks are described in detail, where they are not mentioned fully as a routine maintenance item in the text. If a specific item is mentioned but not described in detail, it will be covered fully in the appropriate Chapter. No special tools are required for the normal routine maintenance tasks. The tools contained in the kit supplied with every machine will prove adequate for each task or if they are not available, the tools found in the average household.

Periodic maintenance intervals

Daily
Check oil level

Every 1,000 miles
Check oil level in primary chaincase
Lubricate throttle linkage

Every 2,000 miles
Check oil level in gearbox
Lubricate rear chain (oil or grease)
Lubricate centre stand and prop stand pivots (oil)
Grease swinging arm
Lubricate exposed cables and joints (oil or grease)
Grease clutch cable
Lubricate brake pedal pivot (oil)
Lubricate ignition advance mechanism (oil)
Lubricate contact breaker cam (oil)
Grease speedometer drive

Every 4,000 miles
Drain and refill the oil reservoir
Clean oil reservoir filter
Replace the oil filter element (if fitted)
Clean crankcase filter and examine pump ball valve
Drain and refill the gearbox
Drain and refill primary chaincase
Grease rear brake cam spindle

Every 10,000 miles
Drain and refill the front forks
Clean and repack wheel bearings with grease
Grease steering head bearings.

Capacities

	B25SS	B25T	B44	B50SS	B50T	B50MX
FUEL TANK:						
gals.	2.5	2.5	1.25	2.5	2.5	1.25
litres	9.0	9.0	7.9	9.0	9.0	4.5
OIL TANK:						
pts.	4.75	4.75	4.8	4.75	4.75	4.75
litres	2.25	2.25	2.3	2.25	2.25	2.25
TRANSMISSION:						
pts.	5/8	5/8	0.6	5/8	5/8	5/8
cc.	280	280	264	280	280	280
PRIMARY CHAINCASE:						
pts.	1/3	1/3	1/3	1/3	1/3	1/3
cc.	140	140	142	140	140	140
FRONT FORKS (per leg):						
pts.	2/5	2/5	2/5	2/5	2/5	2/5
cc.	190	190	190	190	190	190
TYRE PRESSURE:						
front psi.	22	22	16	22	22	—
rear psi.	24	24	17	24	24	—

Torque wrench settings:	lb f ft	(Kg f m)
Clean dry threads		
Carburettor flange nuts	10	(1.383)
Clutch centre nut	60/65	(8.295–8.987)
Con. rod end cap nuts B25/C25	25/27	(3.456–3.733)
Crankpin nuts (B44)	200	(27.65)
Crankshaft pinion nut	35/40	(4.839–5.530)
Cylinder barrel nuts (B25/C25	26/28	(3.595–3.871)
Cylinder barrel nuts (B44)	30/33	(4.148–4.562)
Cylinder head stud nuts	18/20	(2.489–2.765)
Fork leg cap nuts	50/55	(6.913–7.604)
Fork leg pinch bolts	18/20	(2.489–2.765)
Kickstart ratchet nut	50/55	(6.913–7.604)
Oil pump stud nuts	5/7	(.691–.968)
Rotor fixing nut	60	(8.295)
Valve cover nuts	10	(1.383)
Valve cover nuts	5/7	(.691–.968)

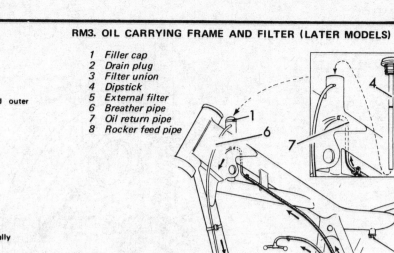

RM3. OIL CARRYING FRAME AND FILTER (LATER MODELS)

1 Filler cap
2 Drain plug
3 Filter union
4 Dipstick
5 External filter
6 Breather pipe
7 Oil return pipe
8 Rocker feed pipe

Nipple

Inner cable

Plasticine funnel around outer cable

Cable suspended vertically

Cable is lubricated when oil drips from far end

RM1. OILING A CONTROL CABLE

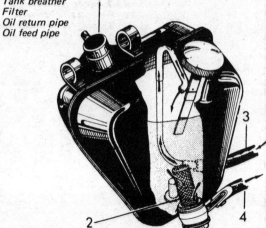

RM4. OIL TANK AND FILTER (EARLIER MACHINES)

1 Tank breather
2 Filter
3 Oil return pipe
4 Oil feed pipe

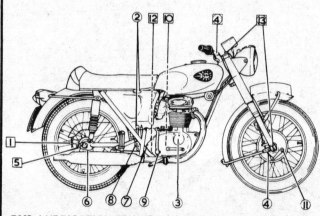

RM2. LUBRICATION POINTS OF A TYPICAL BSA 'SINGLE'

(Numbers in circles refer to right side of machine; numbers in squares refer to left side of machine).

1 Rear chain
2 Oil tank (behind the steering head on later models)
3 Contact breaker cam
4 Control cables
5 Rear brake cam spindle
6 Speedometer cable
7 External oil filter (later models only)
8 Swinging arm pivots
9 Gearbox
10 Primary chaincase
11 Front brake cam spindle
12 Brake pedal pivot
13 Front forks

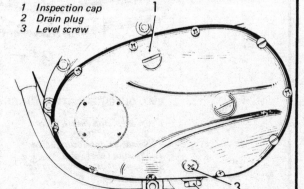

RM5. PRIMARY CHAINCASE

1 Inspection cap
2 Drain plug
3 Level screw

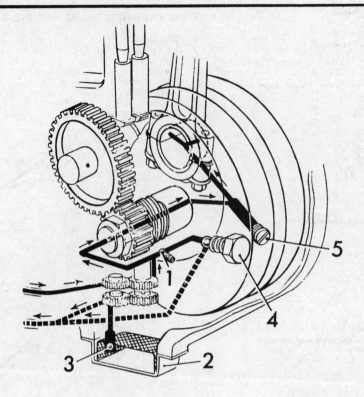

RM6. OIL CIRCULATION DIAGRAM (SHELL TYPE BIG END BEARING)

1 Pressure non-return valve
2 Sump and filter
3 Scavenge non-return valve
4 Pressure relief valve
5 Tapping for oil warning light
  or gauge

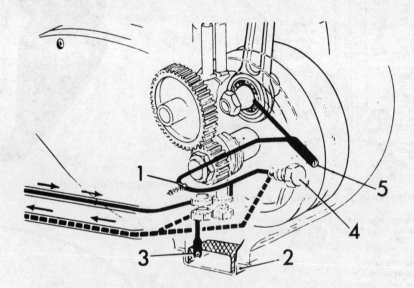

RM7. OIL CIRCULATION DIAGRAM (ROLLER TYPE BIG END BEARING)

1 Pressure non-return valve
2 Sump and filter
3 Scavenge non-return valve
4 Pressure relief valve
5 Tapping for oil warning light
  or gauge

Recommended lubricants

COMPONENT					CORRECT CASTROL PRODUCTS					CAPACITY
ENGINE	...	...	...	...	Castrol GTX	20W/50	...	...	...	4 pints
GEARBOX	...	...	...	...	Castrol Hypoy	90EP	...	...	...	½ pint
					Castrol GTX*	20W/50	...	...	...	¾ pint
PRIMARY CHAINCASE		...	...	...	Castrolite	10W/30	...	...	...	¼ pint
FRONT FORKS	...	...	...	...	Castrolite	10W/30	...	...	...	1/3 pint
					Castrolite*	10W/30	...	...	...	¼ pint
					Castrol TQF+					
REAR CHAIN	...	...	...	...	Castrol Graphited Grease					
WHEEL HUBS										
CONTACT BREAKER CAM										
REAR BRAKE CAM SPINDLE					Castrol LM Grease					
SWINGING FORK PIVOT		...	...	...						
SPEEDOMETER DRIVE										
CLUTCH CABLE										

*C15 and B40 models
+B25 and B50 models

Safety first!

Professional motor mechanics are trained in safe working procedures. However enthusiastic you may be about getting on with the job in hand, do take the time to ensure that your safety is not put at risk. A moment's lack of attention can result in an accident, as can failure to observe certain elementary precautions.

There will always be new ways of having accidents, and the following points do not pretend to be a comprehensive list of all dangers; they are intended rather to make you aware of the risks and to encourage a safety-conscious approach to all work you carry out on your vehicle.

Essential DOs and DON'Ts

DON'T start the engine without first ascertaining that the transmission is in neutral.

DON'T suddenly remove the filler cap from a hot cooling system – cover it with a cloth and release the pressure gradually first, or you may get scalded by escaping coolant.

DON'T attempt to drain oil until you are sure it has cooled sufficiently to avoid scalding you.

DON'T grasp any part of the engine, exhaust or silencer without first ascertaining that it is sufficiently cool to avoid burning you.

DON'T allow brake fluid or antifreeze to contact the machine's paintwork or plastic components.

DON'T syphon toxic liquids such as fuel, brake fluid or antifreeze by mouth, or allow them to remain on your skin.

DON'T inhale dust – it may be injurious to health (see *Asbestos* heading).

DON'T allow any spilt oil or grease to remain on the floor – wipe it up straight away, before someone slips on it.

DON'T use ill-fitting spanners or other tools which may slip and cause injury.

DON'T attempt to lift a heavy component which may be beyond your capability – get assistance.

DON'T rush to finish a job, or take unverified short cuts.

DON'T allow children or animals in or around an unattended vehicle.

DON'T inflate a tyre to a pressure above the recommended maximum. Apart from overstressing the carcase and wheel rim, in extreme cases the tyre may blow off forcibly.

DO ensure that the machine is supported securely at all times. This is especially important when the machine is blocked up to aid wheel or fork removal.

DO take care when attempting to slacken a stubborn nut or bolt. It is generally better to pull on a spanner, rather than push, so that if slippage occurs you fall away from the machine rather than on to it.

DO wear eye protection when using power tools such as drill, sander, bench grinder etc.

DO use a barrier cream on your hands prior to undertaking dirty jobs – it will protect your skin from infection as well as making the dirt easier to remove afterwards; but make sure your hands aren't left slippery. Note that long-term contact with used engine oil can be a health hazard.

DO keep loose clothing (cuffs, tie etc) and long hair well out of the way of moving mechanical parts.

DO remove rings, wristwatch etc, before working on the vehicle – especially the electrical system.

DO keep your work area tidy – it is only too easy to fall over articles left lying around.

DO exercise caution when compressing springs for removal or installation. Ensure that the tension is applied and released in a controlled manner, using suitable tools which preclude the possibility of the spring escaping violently.

DO ensure that any lifting tackle used has a safe working load rating adequate for the job.

DO get someone to check periodically that all is well, when working alone on the vehicle.

DO carry out work in a logical sequence and check that everything is correctly assembled and tightened afterwards.

DO remember that your vehicle's safety affects that of yourself and others. If in doubt on any point, get specialist advice.

IF, in spite of following these precautions, you are unfortunate enough to injure yourself, seek medical attention as soon as possible.

Asbestos

Certain friction, insulating, sealing, and other products – such as brake linings, clutch linings, gaskets, etc – contain asbestos. *Extreme care must be taken to avoid inhalation of dust from such products since it is hazardous to health.* If in doubt, assume that they *do* contain asbestos.

Fire

Remember at all times that petrol (gasoline) is highly flammable. Never smoke, or have any kind of naked flame around, when working on the vehicle. But the risk does not end there – a spark caused by an electrical short-circuit, by two metal surfaces contacting each other, by careless use of tools, or even by static electricity built up in your body under certain conditions, can ignite petrol vapour, which in a confined space is highly explosive.

Always disconnect the battery earth (ground) terminal before working on any part of the fuel or electrical system, and never risk spilling fuel on to a hot engine or exhaust.

It is recommended that a fire extinguisher of a type suitable for fuel and electrical fires is kept handy in the garage or workplace at all times. Never try to extinguish a fuel or electrical fire with water.

Note: *Any reference to a 'torch' appearing in this manual should always be taken to mean a hand-held battery-operated electric lamp or flashlight. It does **not** mean a welding/gas torch or blowlamp.*

Fumes

Certain fumes are highly toxic and can quickly cause unconsciousness and even death if inhaled to any extent. Petrol (gasoline) vapour comes into this category, as do the vapours from certain solvents such as trichloroethylene. Any draining or pouring of such volatile fluids should be done in a well ventilated area.

When using cleaning fluids and solvents, read the instructions carefully. Never use materials from unmarked containers – they may give off poisonous vapours.

Never run the engine of a motor vehicle in an enclosed space such as a garage. Exhaust fumes contain carbon monoxide which is extremely poisonous; if you need to run the engine, always do so in the open air or at least have the rear of the vehicle outside the workplace.

The battery

Never cause a spark, or allow a naked light, near the vehicle's battery. It will normally be giving off a certain amount of hydrogen gas, which is highly explosive.

Always disconnect the battery earth (ground) terminal before working on the fuel or electrical systems.

If possible, loosen the filler plugs or cover when charging the battery from an external source. Do not charge at an excessive rate or the battery may burst.

Take care when topping up and when carrying the battery. The acid electrolyte, even when diluted, is very corrosive and should not be allowed to contact the eyes or skin.

If you ever need to prepare electrolyte yourself, always add the acid slowly to the water, and never the other way round. Protect against splashes by wearing rubber gloves and goggles.

Mains electricity and electrical equipment

When using an electric power tool, inspection light etc, always ensure that the appliance is correctly connected to its plug and that, where necessary, it is properly earthed (grounded). Do not use such appliances in damp conditions and, again, beware of creating a spark or applying excessive heat in the vicinity of fuel or fuel vapour. Also ensure that the appliances meet the relevant national safety standards.

Ignition HT voltage

A severe electric shock can result from touching certain parts of the ignition system, such as the HT leads, when the engine is running or being cranked, particularly if components are damp or the insulation is defective. Where an electronic ignition system is fitted, the HT voltage is much higher and could prove fatal.

Chapter 1 Engine, clutch and gearbox

Contents

Specifications

The BSA range of single cylinder OHV motor cycles manufactured since 1958 all embody the same engine/gearbox unit construction method, and are very similar in design. Most of the photographs in this manual relate to a 250 cc 'Starfire' which was stripped, examined and re-assembled in our workshops while this manual was being compiled. However, where significant variations in design occur between the 250, 350, 441 and 500 cc models the differences are covered by separate paragraphs in the text with additional illustrations where it is considered essential.

250 cc C15 Engine:

Engine bore	67 mm
Engine stroke	70 mm
Engine capacity	249 cc
Oil tank capacity	4 pints
Gearbox capacity	½ pint
Tappet clearance cold:	
Inlet	.008 in
Exhaust	.010 in
Piston ring gap:	
Plain	.010 in
Oil control	.010 in
Piston ring side clearance	.002—.004 in
Piston clearance:	
Bottom of skirt	.0025—.004
Compression ratio	7.25 : 1
Valve spring length:	
C15 and C15T	inner 1 5/8 in, outer 2 1/32 in
C15S and C15 Sport Star	inner 1.5 in, outer 1.67 in

Valve timing - Inlet
    Opens BTDC   ...   .   26 deg.
    Closes ABDC   ...   .   70 deg.

Valve timing - Exhaust
    Opens BTDC   ...   .   61½ deg.
    Closes ATDC   ...   .   34½ deg.

Ignition setting
    Fully advanced   ...   .   11/32 inch before TDC
    Fully retarded   ...   .   —

Gear ratios:
    Top 5.98
    Third 7.65
    Second 10.54
    First 15.96

250 cc B25 Engine:

Cylinder barrel
    Bore diameter (standard)   ...   .   67 mm
    Oversizes   ...   .   0.020 in and 0.040 in   (0.5 and 1.0 mm)
    Material   ...   .   Aluminium alloy LM4 with austenitic iron liner
    Stroke   ...   .   70 mm
    Capacity   ...   .   249 cc

Piston
    Compression ratio   ...   .   10 : 1 (alternative 8.5 : 1)
    Clearance (top of skirt)   ...   .   0.0042—0.0053 in.
    Clearance (bottom of skirt)   ...   .   0.0025—0.0028 in.
    Wrist pin hole diameter   ...   .   0.6884—0.6886 in.
    Material   ...   .   Aluminium alloy H.G. 413

Piston rings
    Width (top and centre)   ...   .   0.101—0.107 in.
    Width (oil control)   ...   .   0.101—0.107 in.
    Depth (top and centre)   ...   .   0.0615—0.0625 in.
    Depth (oil control)   ...   .   0.124—0.125 in.
    Clearance in groove (all rings)   ...   .   0.001—0.003 in.
    Fitted gap (all rings)   ...   .   0.009—0.013 in.
    Material (all rings)   ...   .   Cast iron H.G. 22

Connecting rod and crankshaft assembly
    Connecting rod small end diameter   ...   .   0.6890—0.6894 in.
    Connecting rod big end diameter   ...   .   1.5630—1.5635 in.
    Connecting rod length between centres   ...   .   5.312 in.
    Crankpin diameter   ...   .   1.4375—1.4380 in.
    Regrind undersizes   ...   .   0.010 in, 0.020 in., 0.030 in.   (0.25, o.50, 0.76 mm)
    Journal diameter (left and right)   ...   .   0.9841—0.9844 in.
    Rod bearing running clearance   ...   .   0.0005—0.0015 in.
    Crankcase main bearing (roller, left)   ...   .   0.875 in. x 2.0 in. x 0.5625 in. (Hoffmann R325L)
    Crankcase main bearing (ball, right)   ...   .   0.875 in. x 2.0 in. x 0.5625 in. (Hoffmann 325)

Camshaft
    Journal diameter, left and right   ...   .   0.7480—0.7485 in.
    Cam lift (intake)   ...   .   0.345 in.
    Cam lift (exhaust)   ...   .   0.336 in.
    Base circle radius   ...   .   0.906 in.
    Bush bore diameter, fitted   ...   .   0.7492—0.7497 in.
    Bush outside diameter, left and right hand   ...   ...   0.908—0.909 in.

Tappets
    Stem diameter   ...   .   0.3735—0.3740 in.
    Clearance in crankcase   ...   .   0.001 in.

Cylinder head
    Intake port size   ...   .   1.125 in.
    Exhaust port size   ...   .   1.25 in.
    Material   ...   .   Aluminium alloy LM4 with integral cast iron valve seats

Valves

Seat angle	45°
Head diameter (intake)	1.450–1.455 in.
Head diameter (exhaust)	1.312–1.317 in.
Stem diameter (intake)	0.3095–0.3100 in.
Stem diameter (exhaust)	0.3090–0.3095 in.

Valve guides

Material	Hidural 5
Bore diameter	0.3120–0.3130 in.
Outside diameter	0.5005–0.5010 in.
Length	1.844 in.
Interference fit in head	0.0015–0.0025 in.
Counterbore in exhaust guide	0.323 in.–0.326 in. x 0.12 in. deep

Valve springs

Free length (inner)	1.40 in.	(35.5 mm)
Free length (outer)	1.75 in.	(44.5 mm)
Fitted length (inner)	1.26 in.	(32.0 mm)
Fitted length (outer)	1.37 in.	(34.8 mm)

Clutch

Type	Multiplate with integral cush drive	
Number of friction plates	5	
Number of plain plates	5	
Overall thickness of friction plate	0.167 in.	(4.2 mm)
Free length of springs	1.66 in.	(42.0 mm)
Clutch pushrod length	9.0 in.	(22.9 cm)
Clutch pushrod diameter	0.1875 in.	
Clutch rollers	25, 0.1875 in. x 0.1875 in.	

Gearbox

Type	4 speed, constant mesh
Countershaft bearings (needle roller)	0.5 in. x 0.625 in. x 0.8125 in. (Torrington B108)
Mainshaft bearing (left)	30 x 60 x 16 mm (Hoffmann 130)
Mainshaft bearing (right)	0.625 in. x 1.5625 in. x 0.4735 in. (Hoffmann LS7)
Countershaft diameter (left and right)	0.6245–0.6250 in.
Mainshaft diameter (left)	0.7485–0.7490 in.
Mainshaft diameter (right)	0.6245–0.6250 in.
Sleeve pinion inside diameter	0.752–0.753 in.
Sleeve pinion outside diameter	1.179–1.180 in.

Gear ratios

Gearbox:	B25SS, B25T
Top	1.00
Third	1.24
Second	1.64
First	2.65

Chains

Primary chain (all models)

Pitch	0.375 in.	(9.53 mm)
Roller diameter	0.250 in.	(6.35 mm)
Distance between plates	0.225 in.	(5.72 mm)
Length	70 links	
Breaking load	3,900 lbs	(1770 kg)
Type	Renolds 114 038 Duplex endless	

350 cc B40 Engine

Capacity	343 cc
Cylinder bore	79 mm
Stroke	70 mm
Compression ratio	7 : 1
Inlet opens BTDC	26°
Inlet closes ABDC	70°
Exhaust opens BBDC	61½°
Exhaust closes ATDC	34½°
Piston rings - compression	.0625 wide (in.)
Piston rings - scraper	.125 wide (in.)
Piston rings gaps - minimum	.009 in.
- maximum	.014 in.
Plug points gap - minimum	.020 in.
- maximum	.025 in.

Transmission

Gear ratios	- Top	...	...	...	...	...	...	5.48
	- Third	...	...	...	...	...	...	7.0
	- Second	...	...	...	...	...	...	9.63
	- First	...	...	...	...	...	...	14.6

Teeth on - engine sprocket	...	...	...	...	...	23T
gearbox sprocket	...	...	...	...	...	19T
clutch sprocket	...	...	...	...	...	52T
rear chainwheel	...	...	...	...	...	46T

Clutch friction plates	...	...	...	...	...	...	4
Chain sizes - Front (in.)	...	...	...	...	...	3/8 Duplex (70 pitches)

Capacities

Oil tank	...	...	...	...	...	...	...	4 Imp. pints (2¼ litres)
Gearbox	...	...	...	...	...	...	...	1/5 Imp. pints (220 cc)
Primary chaincase	...	...	...	...	...	...	1/4 Imp. pints (140 cc)

441 cc B44 Engine

Cylinder barrel

Material	...	...	...	...	...	...	Aluminium with austenitic iron liner
Bore size (standard)	...	...	...	...	...	...	79 mm
Stroke	...	...	...	...	...	...	90 mm
Oversizes	...	...	...	...	...	...	0.010 in. and 0.020 in. (0.254 and 0.508 mm)
Capacity	...	...	...	...	...	...	441 cc

Piston

Material	...	...	...	...	...	...	'Lo-Ex' aluminium
Compression ratio	...	...					9.4 : 1
Clearance (bottom of skirt)	...	...	...	...	...	0.003–0.0035 in. (0.0762–0.0889 mm)
Clearance (top of skirt)	...	...	...	...	...	0.006–0.0065 in. (0.0524–0.1651 mm)

(both measured on major axis).

Piston rings

Material - compression (top)	...	...	...	...	...	Brico 8
Material - compression (top)	...	...	...	...	...	Brico BSS.5004 (chrome-plated)
Material - compression (centre)	...	...	...	...	...	Brico 8
Material - scraper	...	...	...	...	...	...	Brico BSS.5004
Width - compression (top and centre)	...	...	...	0.0625 in. (1.5875 mm)
Width - scraper	...	...	...	...	...	...	0.125 in. (3.175 mm)
Depth	...	...	...	...	...	...	0.120–0.127 in. (3.048–3.2258 mm)
Clearance in groove	...	...	...	...	...	0.001–0.003 in. (0.0254–0.0762 mm)
Fitted gap - (maximum)	...	...	...	...	...	0.014 in. (0.3556 mm)
Fitted gap - (minimum)	...	...	...	...	...	0.009 in. (0.2283 mm)

Oil pump

Pump body material	...	...	...	...	...	...	Zinc base alloy
Type	...	...	...	...	...	...	...	Double gear
Drive ratio	...	...	...	...	...	...	1 : 1
Non-return valve spring (free length)	...	...	...	0.5 in. (12.7 mm)
Non-return valve spring ball (diameter)	...	...	0.25 in. (6.35 mm)
Oil pressure release valve spring (free length)	...	0.6094 in. (15.4781 mm)
Oil pressure release valve ball (diameter)	...	...	...	0.3125 in. (7.9375 mm)

Bearing dimensions

Clutch roller (25)	...	...	...	...	...	0.1875 x 0.1875 in. (4.7025 x 4.7025 mm)
Con-rod big-end bush (bore)	...	...	...	...	1.7701–1.7706 in. (44.9605–44.9732 mm)
							0.250 in. dia. x
Con-rod big-end roller (24)	...	...	...	...	...	0.250 in. (6.35 x 6.35 mm)
Con-rod small-end bush (bore)	...	...	...	...	0.7503–0.7506 in. (19.0576–19.0652 mm)
Crankpin diameter	...	...	...	...	...	1.2698–1.2700 in. (32.253–32.258 mm)
Crankcase bearing (drive-side)	...	...	...	...	25 x 62 x 17 mm
Crankcase bearing (gear-side)	...	...	...	...	25 x 62 x 17 mm
Flywheel shaft diameter (drive-side and gear-side)	...	0.9841–0.9844 in. (24.9961–25.0038 mm)
Gearbox layshaft bearings (drive-side and gear-side)	...	0.5 x 0.625 x 0.8125 in. (12.7 x 15.875 x 20.6375 mm)
Gearbox layshaft diameter (drive-side and gear-side)	...	0.6245–0.625 in. (15.8623–15.8750 mm)
Gearbox mainshaft bearing (drive-side)	...	...	...	30 x 62 x 16 mm
							0.625 x 1.5625 x (15.875 x 39.2875 x
Gearbox mainshaft bearing (gear-side)	...	...	...	0.4375 in. 11.1125 mm)
Gearbox mainshaft diameter (drive-side)	...	...	...	0.7485–0.749 in. (19.0119–19.0246 mm)
Gearbox mainshaft diameter (gear-side)	...	...	...	0.6245–0.625 in. (15.8623–15.8750 mm)

Gearbox sleeve pinion (internal diameter)	0.752–0.753 in.	(19.1008–19.1262 mm)
Gearbox sleeve pinion (external diameter)	1.179–1.180 in.	(29.9466–29.9720 mm)
Gudgeon pin diameter	0.750–0.7502 in.	(19.05–19.055 mm)

Camshaft

Journal diameter (left-hand)	0.5598–0.5603 in.	(14.2189–14.2316 mm)
Journal diameter (right-hand)	0.7480–0.7485 in.	(18.9992–19.0119 mm)
Cam lift (inlet)	0.345 in.	(8.763 mm)
Cam lift (exhaust)	0.336 in.	(8.534 mm)
Base circle radius	0.386 in.	(9.8044 mm)

Camshaft bearing bushings

Bore diameter (fitted) left-hand	0.561–0.562 in.	(14.2494–14.2748 mm)
Bore diameter (fitted) right-hand	0.7492–0.7497 in.	(19.0297–19.04238 mm)
Outside diameter (left-hand)	0.719–0.720 in.	(18.2626–18.2880 mm)
Outside diameter (right-hand)	0.908–0.909 in.	(23.0632–23.0886 mm)
Camshaft clearance (left-hand)	0.0007–0.0022 in.	(0.01778–0.05588 mm)
Camshaft clearance (right-hand)	0.0007–0.0017 in.	(0.01778–0.04318 mm)

* *Denotes 1968 model.*

Cylinder head

Material	Aluminium alloy	
Inlet port size	1.125 in.	(28.575 mm)
Exhaust port size	1.25 in.	(31.75 mm)

Valves

Seat angle (inclusive)	90º	
Head diameter (inlet)	1.535–1.540 in.	(38.9890–39.1160 mm)
Head diameter (exhaust)	1.407–1.412 in.	(35.737–35.864 mm)
Stem diameter (inlet)	0.3095–0.3100 in.	(7.861–7.874 mm)
Stem diameter (exhaust)	0.3090–0.3095 in.	(7.848–7.861 mm)

Valve guides

Material	Phosphor bronze	
Bore diameter	0.3120–0.3130 in.	(7.9248–7.950 mm)
Outside diameter	0.5005–0.5010 in.	(12.7127–12.7254 mm)
Length	1.859 in.	(47.2186 mm)
Cylinder head interference fit	0.0015–0.0025 in.	(0.0381–0.0635 mm)

Valve springs

Free length (inner)	1.500 in.	(38.10 mm)
Free length (outer)	1.670 in.	(42.418 mm)
Fitted length (inner)	1.218 in.	(30.9372 mm)
Fitted length (outer)	1.312 in.	(33.3248 mm)

Clutch

Type	Multi-plate with integral cush drive	
Number of plates:		
Driving (bonded segments)	4	
Driven (plain)	5	
Overall thickness of driving plate and segments	0.167 in.	(4.242 mm)
Clutch springs	4	
Free length of springs	1.65685 in.	(42.0687 mm)
Clutch push rod (length)	9.0 in.	(228.6 mm)
Clutch push rod (diameter)	0.1875 in.	(4.7025 mm)

Gear ratios

Gearbox: - (top)	1.0	
(third)	1.24	
(second)	1.65	
(first)	2.65	
Overall - (top)	5.14	
(third)	6.39	
(second)	8.45	
(first)	13.62	

Sprockets

Engine	28 teeth	
Clutch	52 teeth	
Gearbox	17 teeth	

Chain sizes:
    Primary ... ...   Duplex 0.375 in. x 72 links

B50 Engine

Cylinder barrel
    Bore diameter (standard) ... ...   84 mm
    Oversizes ... ...   0.020 in. and 0.040 in.   (0.5 mm and 1.0 mm)
    Material ... ...   Aluminium alloy LM4M with austenitic iron liner
    Stroke ... ...   90 mm
    Capacity ... ...   499 cc

Piston
    Compression ratio ... ...   10 : 1
    Clearance (top of skirt) ... ...   0.005—0.007 in.
    Clearance (bottom of skirt) ... ...   0.0035—0.0045 in.
    Wrist pin hole diameter ... ...   0.7499—0.7501 in.
    Material ... ...   Aluminium alloy H.G. 413

Piston rings
    Width (top and centre) ... ...   0.127—0.134 in.
    Width (oil control) ... ...   0.138—0.145 in.
    Depth (top and centre) ... ...   0.0615—0.0625 in.
    Depth (total, oil control) ... ...   0.1550—0.1560 in.
    Clearance in groove (all rings) ... ...   0.001—0.003 in.
    Fitted gap (all rings) ... ...   0.016—0.024 in.   (0.40—0.60 mm)
    Material (all rings) ... ...   Cast iron H.G. 22

Connecting rod and crankshaft assembly
    Connecting rod small end diameter ... ...   0.8115—0.8125 in.
    Small end bush bore diameter ... ...   0.7503—0.7506 in.
    Connecting rod big end diameter ... ...   2.0190—2.0195 in.
    Big end bush bore diameter ... ...   1.8110—1.8116 in.
    Connecting rod length between centres ... ...   6.00 in.   (15.24 cm)
    Crankpin bearing diameter ... ...   1.4957—1.4961 in.
    Big end bearing (needle roller) ... ...   38 x 46 x 20 mm (R. & M K38-46-20F)
    Flywheel shaft diameter (left and right) ... ...   0.9841—0.9844 in.
    Crankcase main bearings (roller, left and right) ... ...   0.875 in. x 2.0 in. x 0.5625 in. (Hoffmann R325L)
    Crankcase main bearing (ball, left) ... ...   0.875 in. x 2.0 in. x 0.5625 in. (Hoffmann LS9)

Camshaft
    Journal diameter, left to right ... ...   0.7480—0.7485 in.
    Cam lift (intake) ... ...   0.345 (0.355 in. B50MX)
    Cam lift (exhaust) ... ...   0.336 (0.355 in. B50MX)
    Base circle radius ... ...   0.906 in.
    Bush bore diameter, fitted ... ...   0.7492—0.7497 in.
    Bush outside diameter, left and right ... ...   0.908—0.909 in.
    Tappet clearance (intake) ... ...   0.008 in.   (0.20 mm)
    Tappet clearance (exhaust) ... ...   0.010 in.   (0.25 mm)

Cylinder head
    Intake port size ... ...   1.20 in.
    Exhaust port size ... ...   1.625 in.
    Material ... ...   Aluminium alloy LM4 with cast iron valve seats

Valves
    Seat angle ... ...   45°
    Head diameter (intake) ... ...   1.750—1.755 in.
    Head diameter (exhaust) ... ...   1.526—1.531 in.
    Stem diameter (intake) ... ...   0.3100—0.3105 in.
    Stem diameter (exhaust) ... ...   0.3095—0.3100 in.

Valve guides
    Material ... ...   Phosphor-bronze
    Bore diameter ... ...   0.3120—0.3130 in.
    Outside diameter ... ...   0.5005—0.5010 in.
    Length ... ...   1.859 in.
    Interference fit in head ... ...   0.0015—0.0025 in.
    Counterbore in exhaust guide ... ...   0.323—0.326 in. x 0.12 in. deep

Valve springs
    Free length (inner) ... ...   1.50 in.   (38.1 mm)
    Free length (outer) ... ...   1.67 in.   (42.4 mm)

Fitted length (inner)	...	...	...	...	...	...	...	1.22 in.	(31.0 mm)	
Fitted length (outer)	...	...	...	...	...	...	...	1.31 in.	(33.3 mm)	

Valve timing
Tappets set to 0.015 in. (0.38 mm)

								B50SS, B50T	B50MX
for checking purposes only:—	...	...	...	...	...	...			
Intake opens BTDC	...	...	...	...	...	...		51°	63°
Intake closes ABDC	...	...	...	...	...	...		68°	72°
Exhaust opens BBDC	...	...	...	...	...	...		78°	80°
Exhaust closes ATDC	...	...	...	...	...	...		37°	55°

Clutch

Type	...	...	...	...	...	...	...	...	Multiplate with integral cush drive	
Number of discs	...	...	...	...	...	...	...	5		
Number of plates	...	...	...	...	...	...	...	5		
Overall thickness of friction plate	...	...	...	...	...	0.167 in.	(4.2 mm)	
Free length of springs	...	...	...	...	...	...	1.66 in.	(42.0 mm)	
Clutch pushrod length	...	...	...	...	...	...	9.0 in.	(22.9 cm)	
Clutch pushrod diameter	...	...	...	...	...	...	0.1875 in.		
Clutch rollers	...	...	...	...	...	...	...	0.1875 in. x 0.1875 in., 25	

Gearbox

Type	...	...	...	...	...	...	4 speed, constant mesh
Countershaft bearings (needle roller)	...	...	...	0.5 in. x 0.625 in. x 0.8125 in. (Torrington B108)
Mainshaft bearing (left)	...	...	...	...	...	30 x 60 x 16 mm (Hoffmann 130)
Mainshaft bearing (right)	...	...	...	...	...	0.625 in. x 1.5625 in. x 0.4735 in. (Hoffmann LS7)
Countershaft diameter (left and right)	...	...	...	0.6245—0.6250 in.
Mainshaft diameter (left)	...	...	...	...	0.7485—0.7490 in.
Mainshaft diameter (right)	...	...	...	...	0.6245—0.6250 in.
Sleeve pinion inside diameter	...	...	...	...	0.752—0.753 in.
Sleeve pinion outside diameter	...	...	...	...	1.179 in. x 1.180 in.

Chains
Primary chain (all models)

Pitch	...	...	...	...	...	...	...	0.375 in.	(9.53 mm)	
Roller diameter	...	...	...	...	...	...	0.250 in.	(6.35 mm)	
Distance between plates	...	...	...	...	...	0.225 in.	(5.72 mm)	
Length	...	...	...	...	...	...	...	72 links		
Breaking load	...	...	...	...	...	...	3,900 lbs	(1770 kg)	
Type	...	...	...	...	...	...	...	Renolds 114 038 Duplex endless	

1 General description

The 250,350, 441 and 500 cc BSA engines are all unit construction, overhead valve singles. The earlier 250 and 350 cc Star models are virtually identical having roller bearing big-end journals, screw and locknut tappet adjusters and a plain crankshaft bearing on the timing case side of the engine. The later 250 cc Starfire models differ by having a plain shell-type big end bearing, a ball race on both sides of the crankshaft, rotary cam adjusted tappets and larger, square section cooling fins.

The 441 cc B44 and 500 cc and B50 engines are also very similar in design and have reverted back to the use of roller bearing big ends and screw adjusted tappets. The B50 has additional bearing on the primary drive side of the crankshaft to cater for the extra power and a steel connecting rod instead of an alloy one for the same reason. All engines are fitted with aluminum alloy pistons with three rings. The C15 and B40 models have cast iron cylinder heads while all subsequent engines have aluminum heads.

The left hand side primary drive case houses the clutch assembly, alternator and primary chain. The rear section of the crankcase casting encloses a four speed, constant mesh gearbox which has its own oil bath and is independent of the engine lubrication system. The heart of the engine lubrication system is a two stage, gear type oil pump driven from the timing side of the crankshaft. Oil is pressure fed to the big end bearing and, on ealier 250 and 350 cc models, to the plain crankshaft bearing. Centrifugal force distributes oil from the big end journal to the other moving parts of the engine. The rocker gear is fed by an external pipe at a lower pressure.

Owners of Triumph 250 Trophy machines will also find this Chapter extremely useful as these engines and that of the 250 Starfire are virtually identical.

2 Operations with engine in frame

It is not necessary to remove the engine unit from the frame unless the crankshaft assembly or main bearings require attention. Most operations can be accomplished with the engine in place, such as:
1 Removal and replacement of cylinder head.
2 Removal and replacement of cylinder barrel and piston.
3 Removal and replacement of flywheel generator.
4 Removal and replacement of clutch assembly.
5 Removal and replacement of contact breaker assembly.
6 Removal and replacement of kickstart quadrant assembly.
7 Removal and replacement of the gear cluster, selectors and gearbox main bearings.

When several operations need to be undertaken simultaneously, it will probably be avantageous to remove the complete engine unit from the frame, an operation that should take approximately one hour. This will give the advantage of better access and more working space.

3 Operations with engine removed

1 Removal and replacement of the main bearings.
2 Removal and replacement of the crankshaft assembly.

4 Method of engine/gearbox removal

As described previously, the engine and gearbox are built in unit and it is necessary to remove the unit complete, in order to gain access to the crankshaft assembly. Separation is accomplished after the engine unit has been removed and refitting

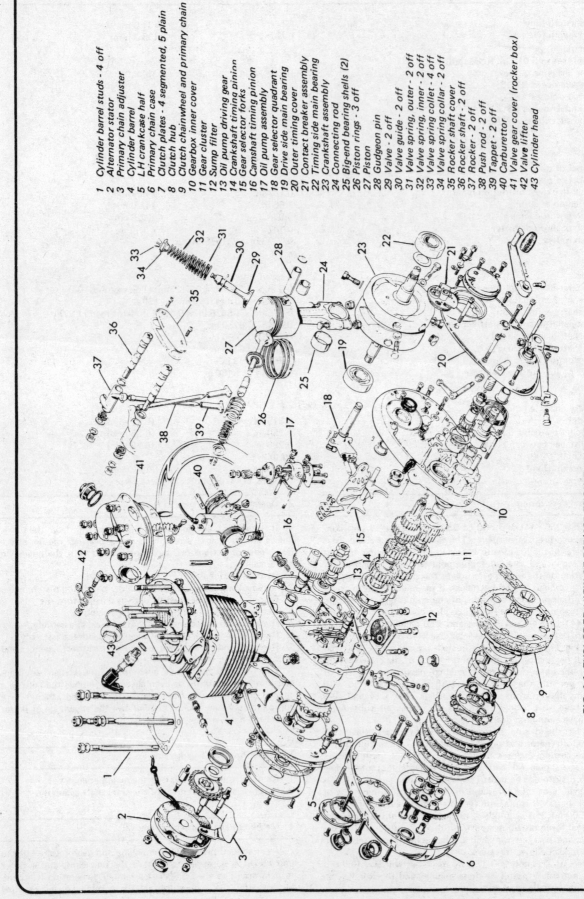

1 Cylinder barrel studs - 4 off
2 Alternator stator
3 Primary chain adjuster
4 Cylinder barrel
5 LH crankcase half
6 Primary chain case
7 Clutch plates - 4 segmented, 5 plain
8 Clutch hub
9 Clutch chainwheel and primary chain
10 Gearbox inner cover
11 Gear cluster
12 Sump filter
13 Oil pump driving gear
14 Crankshaft timing pinion
15 Gear selector forks
16 Camshaft timing pinion
17 Oil pump assembly
18 Gear selector quadrant
19 Drive side main bearing
20 Outer timing cover
21 Contact breaker assembly
22 Timing side main bearing
23 Crankshaft assembly
24 Connecting rod
25 Big-end bearing shells (2)
26 Piston rings - 3 off
27 Piston
28 Gudgeon pin
29 Valve - 2 off
30 Valve guide - 2 off
31 Valve spring, outer - 2 off
32 Valve spring, inner - 2 off
33 Valve spring collet - 4 off
34 Valve spring collar - 2 off
35 Rocker shaft cover
36 Rocker shaft - 2 off
37 Rocker - 2 off
38 Push rod - 2 off
39 Tappet - 2 off
40 Carburettor
41 Valve gear cover (rocker box)
42 Valve lifter
43 Cylinder head

FIG. 1.1. EXPLODED VIEW OF B25/C25 ENGINE AND GEARBOX

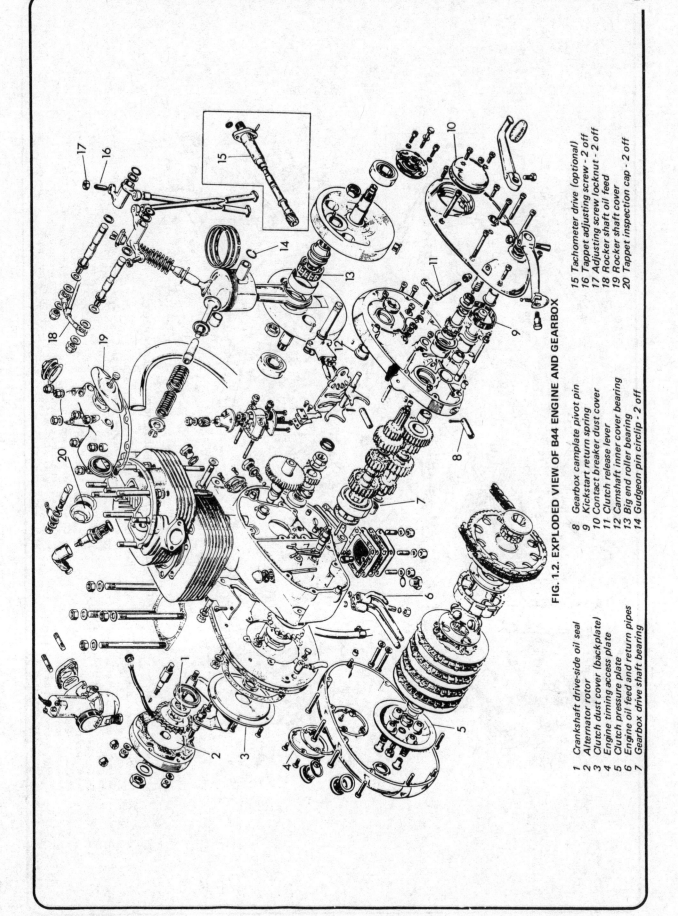

FIG. 1.2. EXPLODED VIEW OF B44 ENGINE AND GEARBOX

1 Crankshaft drive-side oil seal
2 Alternator rotor
3 Clutch dust cover (backplate)
4 Engine timing access plate
5 Clutch pressure plate
6 Engine oil feed and return pipes
7 Gearbox drive shaft bearing

8 Gearbox camplate pivot pin
9 Kickstart return spring
10 Contact breaker dust cover
11 Clutch release lever
12 Camshaft inner cover bearing
13 Big end roller bearing
14 Gudgeon pin circlip - 2 off

15 Tachometer drive (optional)
16 Tappet adjusting screw - 2 off
17 Adjusting screw locknut - 2 off
18 Rocker shaft oil feed
19 Rocker shaft cover
20 Tappet inspection cap - 2 off

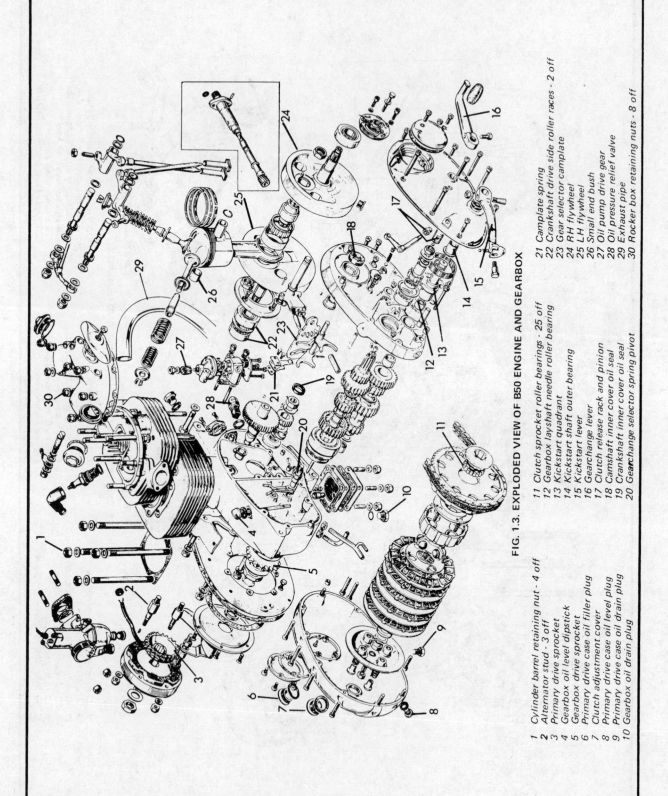

FIG. 1.3. EXPLODED VIEW OF B50 ENGINE AND GEARBOX

1 Cylinder barrel retaining nut - 4 off
2 Alternator stud - 3 off
3 Primary drive sprocket
4 Gearbox oil level dipstick
5 Gearbox drive sprocket
6 Primary drive case oil filler plug
7 Clutch adjustment cover
8 Primary drive case oil level plug
9 Primary drive case oil drain plug
10 Gearbox oil drain plug

11 Clutch sprocket roller bearings - 25 off
12 Gearbox layshaft needle roller bearing
13 Kickstart quadrant
14 Kickstart shaft outer bearing
15 Kickstart lever
16 Gearchange lever
17 Clutch release rack and pinion
18 Camshaft inner cover oil seal
19 Crankshaft inner cover oil seal
20 Gearchange selector spring pivot

21 Camplate spring
22 Crankshaft drive side roller races - 2 off
23 Gear selector camplate
24 RH flywheel
25 LH flywheel
26 Small end bush
27 Oil pump drive gear
28 Oil pressure relief valve
29 Exhaust pipe
30 Rocker box retaining nuts - 8 off

cannot take place until the crankcases have been reassembled. When the crankcases are separated the gearbox internals will also be exposed.

5 Removing the engine/gearbox unit

1 Place the machine on the centre stand and make sure it is standing firmly on level ground.
2 Turn off the fuel supply tap and disconnect the fuel pipe from the carburettor.
3 Remove rubber blanking grommet from the centre of the tank, unfasten the retaining nut from its stud and remove tank.
4 Disconnect the CB and SW leads from the ignition coil and remove coil and HT lead (make a note of the colour coding on the LT leads before removing).
5 Remove engine steadying bar and compression release cable if fitted.
6 Remove carburettor and air cleaner from the intake flange, unfasten throttle slide from carburettor and tie the control cables out of the way.
7 Remove rocker feed oil pipe. On some engines the oil feed is led to the rocker shafts via two pipes secured with domed nuts.
8 Disconnect the two footrests and unfasten the nuts and bolts at the two exhaust system attachment points. Slacken the finned exhaust collar and remove the silencer first, then the front pipe.
9 Disconnect the oil pressure feed pipe and tachometer drive if fitted.
10 Unfasten the electrical wiring harness at the snap connectors making a note of the colour coding, detach the clutch operating cable and tie out of the way.
11 Detach engine oil feed and return pipes noting which pipe goes onto which union and allow oil to drain into suitable container. On B50 engines the oil can be drained faster by removing the union at the bottom of the front downtube.
12 Remove the rear chain by means of the disconnecting link. Unfasten the two chainguard attachment bolts and withdraw guard from the rear of the machine.
13 The engine is held in the frame by two short studs, one at the front and one at the rear, and a long stud underneath the crankcase. Remove all three studs and lift the engine/gearbox from the frame. On some engines it may be necessary to first remove an engine plate adjacent to the rear mounting point.
14 Remove the kickstarter and gearchange lever by slackening the pinch bolts.

5.3b Lift the rear of the tank first

5.4 Remove ignition coil and HT lead

5.3a Fuel tank retaining bolt

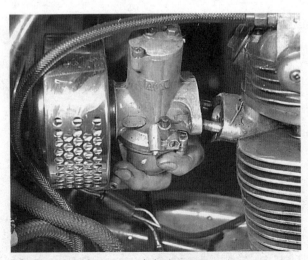
5.6a Remove carburettor and air cleaner

5.6b Disconnect throttle slide

5.11 Engine oil feed and return pipes

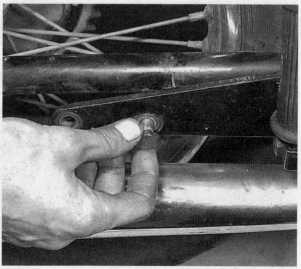

5.8 Silencer attachment point

5.12 Withdraw chainguard towards rear of machine

5.9 Method of clutch cable removal

5.13a Front engine attachment point

5.13b Rear engine attachment point

6 Dismantling the engine, clutch and gearbox - general

Before commencing work on the engine unit, the external surfaces should be cleaned thoroughly. A motor cycle engine has very little protection from road grit and other foreign matter, which will find its way into the dismantled engine if this simple precaution is not observed. One of the proprietary cleaning compounds such as 'Gunk' can be used to good effect, particularly if the compound is allowed to work into the film of oil and grease before it is washed away. When washing down, make sure that water cannot enter the carburettor or the electrical system, particularly if these parts have been exposed.

Never use undue force to remove any stubborn part, unless mention is made of this requirement. There is invariably good reason why a part is difficult to remove, often because the dismantling operation has been tackled in the wrong sequence. Dismantling will be made easier if a simple engine stand is constructed that will correspond with the engine mounting points. This arrangement will permit the complete unit to be clamped rigidly to the work bench, leaving both hands free.

7 Dismantling the engine, clutch and gearbox - removing the rocker gear, cylinder head, barrel and piston

1   Remove the two rocker inspection caps and rotate the engine until both valves are closed and there is clearance at the rocker arms.

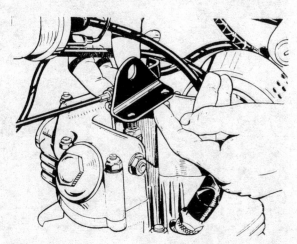

Fig. 1.4 Removing the cylinder head steady assembly

2   Remove the engine steady bracket and nine retaining nuts and lift off the rocker assembly.
3   Remove the two pushrods.
4   Unfasten the cylinder head nuts; some of these are rather awkward to get at, the best tool is a flat combination spanner.
5   Gently slide the cylinder head off the studs.
6   Slacken the two nuts just below the cylinder barrel on the left hand side of the engine and carefully slide the cylinder barrel up the crankcase studs.
7   Remove one of the gudgeon pin circlips using a small screwdriver or long nosed pliers. Push out the gudgeon pin and remove piston complete with rings. If only a 'top' overhaul is contemplated at this time, it is advisable to pad the mouth of the crankcase with rag, to prevent the mishap of a displaced circlip falling in.
8   If the gudgeon pin is a tight fit and cannot be withdrawn easily, warm the piston with a rag soaked in hot water. This will expand the piston bosses sufficiently to release their hold on the pin. Before the pin is withdrawn, make sure the crankcase mouth is covered, to obviate the risk of broken piston rings falling in. Remove the piston complete.
9   Mark the piston INSIDE the skirt so that it is replaced in the same position.

8 Clutch, alternator and primary chain - dismantling and removal

1   If the engine is still in the frame it will be necessary to remove the left side footrest and brake pedal.
2   Remove the drain screw and drain the oil from the primary chaincase. Unfasten the screws and take off the primary drive cover. It may be necessary to break the joint by tapping gently with a rawhide mallet, place a wide drip tray underneath to catch any residual oil.
3   Remove the lockwire (B50 only) from the four clutch spring retaining nuts and withdraw the pressure plate, springs and cups.
4   Withdraw the clutch plates noting the correct order of assembly.
5   Prevent the clutch from turning by jamming the secondary drive sprocket with a piece of wood and remove the clutch centre nut after first bending back the tab washer.
6   Remove the tab washer, spacer and clutch pushrod.
7   To remove the complete clutch assembly and primary chain the alternator will also have to be removed. Unfasten the three stator retaining nuts, pull the alternator lead through the grommet and slide the stator off the studs.
8   Remove the primary chain tensioner noting that a spacer is installed on the rear stud.
9   Bend back the locktab and unscrew the engine shaft nut. Remove the rotor, wipe it clean, and store it in a clean place. Do not lose the woodruff key.
10 Withdraw the clutch centre boss. The clutch outer drum and the engine sprocket must be removed simultaneously, still meshed with the primary chain. Be prepared to catch the twenty five clutch bearing rollers as the drum is displaced.
11 The clutch centre sleeve is located on the mainshaft end by a taper and woodruff key, and is invariably a very tight fit. To aid removal the sleeve is threaded internally to accept a puller (BSA extractor No 61-3583). If this extractor is not available a small two-legged puller may be used to accomplish removal.
12 Before removing the crankshaft distance piece, where fitted, note in which direction it's chamfered edge faces. Make a note of this as a guide to reassembly.

9 Gearbox sprocket - removal

1   Remove the six screws retaining the clutch backplate and remove complete with oil seal.
2   Jam the sprocket by means of a wooden wedge; bend back the tab washer and unscrew the retaining nut. Pull the sprocket off the gearbox shafts.

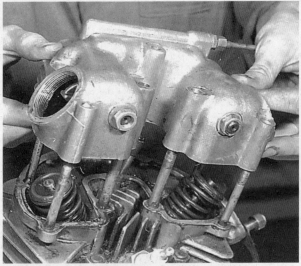

7.2 Removing rocker cover

7.4 Use a flat combination spanner for these cylinder head nuts

7.6 Lifting the cylinder barrel off the crankcase studs

7.7 Removing piston complete with rings

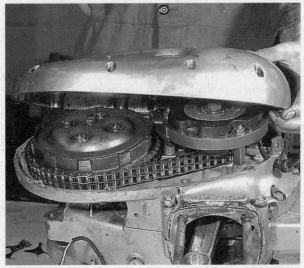

8.2 Removing primary drive cover

8.3 Unscrew clutch springs

8.4 Note correct order of clutch plates

8.7 Remove the alternator stator ...

8.9 ... and rotor

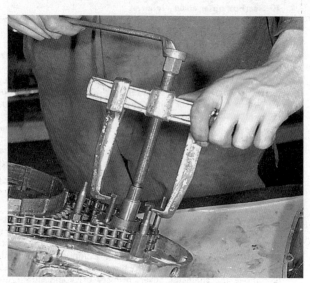

8.10a Using a puller to release the engine drive sprocket

8.10b Removing the primary chain

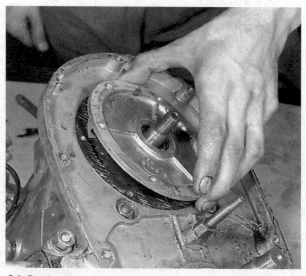

9.1 Gearbox sprocket access plate

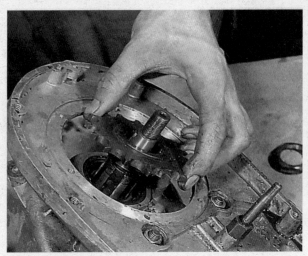

9.2 Removing gearbox sprocket

10 Gearbox outer cover - removal

1   Remove the contact breaker cover, detach the LT wire and withdraw the contact breaker mechanism.
2   If the engine is still in the frame, take off the kickstart and gearchange levers.
3   Remove the outer cover retaining screws noting that some are different lengths and must be replaced in the correct positions.
4   Withdraw the cover taking care not to lose the clutch release rack and ball.
5   Detach the single retaining bolt from the timing shaft and pull the ignition advance unit off its taper. Note that the shaft is threaded to take the BSA extractor tool, which will withdraw the ATU without damaging it; a $5/16$ in UNF bolt can be used if the extractor tool is not available.

11 Kickstart and gearbox components - removal

1   The gearbox inner cover is withdrawn with the gearbox components attached. Resist any temptation to remove the cover, leaving the gearbox components in position, as this approach will make reassembly virtually impossible. Note that if the unit is still in position in the frame, it will be necessary to remove the chaincase cover, primary drive and clutch, gearbox sprocket access plate and the gearbox sprocket. This will allow the gearbox mainshaft to be withdrawn. See Sections 8 and 9 for details.
2   Unscrew the kickstart return spring anchoring nut and remove the return spring. The kickstart mechanism should now be removed. Pull the kickstart quadrant from its bearing, and slide the pinion and ratchet assembly, complete with sleeve, off its shaft, having first removed the securing nut. It is essential to remove the aforementioned components to allow the bearing to be drawn off the splined shaft as the cover is withdrawn.
3   Take out the remaining inner cover mounting screws and tap the cover with a rubber mallet to break the joint seal. Withdraw the cover, complete with transmission gear cluster. As the cover is removed exert a slight inward pressure on the end of the camshaft to avoid disturbing the valve timing of the engine.
4   Insert a suitable flat blade between the camplate and gearchange quadrant, thus depressing the plungers, and withdraw the gearchange quadrant complete with spring.
5   The spring loaded plungers are retained by a small plate, secured with one screw.
6   The gearchange return spring pivot bolt need not be disturbed.
7   Take out the cam plate pivot retaining split pin from the outside of the cover, and, using one of the small inner timing cover screws as an extractor, pull out the pivot with a pair of pliers.

The cam plate may now be taken out, together with the selector forks and their spindle, permitting removal of the layshaft complete with gears and the mainshaft sliding gear. Note that although the selector forks are of similar dimensions, the mainshaft fork has a turned shoulder for identification purposes. It is most important that the forks are correctly replaced.
8   The mainshaft fixed gears and the kickstart ratchet assembly remain fixed to the inner cover bearing; to remove take off kickstart ratchet retaining nut and remove ratchet components from the shaft.
Note: On B50 models the kickstart ratchet assembly is fitted in a reverse order to the B44, B25 and earlier models and there is an additional washer which is fitted against the bearing, followed by the ratchet pinion bush and spring.
9   The two gears remaining on the mainshaft are an interference fit. Remove by clamping the gears in a vice fitted with soft clamps, using cloth to protect the gear teeth from damage, and drive the shaft out with an aluminum drift.
10 To remove the left hand gearbox bearing from the case, drive the pinion out using a hide mallet and remove the oil seal. Before attempting to drive out the bearing, heat the crankcase gently and evenly with a blowlamp or gas torch.
Note: On earlier 250 and 350 Star engines the gear cluster assembly cannot be withdrawn with the inner cover. After the kickstart assembly has been removed, take off the small plate at the rear of the gearchange shaft which covers the pivot for the cam plate. Withdraw the split pin and remove the pivot pin rearwards. The cam shaft tab washer and nut must also be removed (vertical contact breaker engines), together with thrust washer and locating peg. After removing the retaining screws the cover can now be taken off. Before the gear cluster can be drawn out the selector quadrant spring pivot pin must be unscrewed.

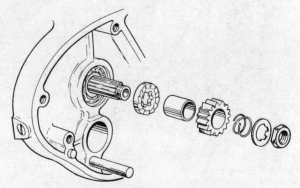

Fig. 1.5. Kickstart ratchet assembly (B25 and B44)

10.1 Removing contact breaker assembly

10.4 Outer cover removed

11.2 Kickstart spring and gear mechanism

12 Oil pump - removal

All models after 1966:

1 Before removing the oil pump, note the timing marks on the camshaft and crankshaft pinions and then withdraw the camshaft and two tappets.

2 Bend back the tab washer and remove the crankshaft nut which has a normal RH thread.

3 Remove the crankshaft pinion using a suitable puller.

Note: This pinion is rather difficult to grip with the ordinary type of puller and if obtainable BSA tool No 61-3773 should be used. Another method of removal is shown in the photograph adjacent, but make sure a small bolt is placed inside the crankshaft end to avoid damage.

4 Take off the three nuts and remove the oil pump. Slide off the pump driving gear and thrust washer taking care not to lose the woodruff key.

All models prior to 1966:

5 On the earlier 250 and 350 Star engine, the contact breaker unit and oil pump were driven from the crankshaft pinion by a common shaft. Before the pump can be removed, the contact breaker unit must be drawn out from the rear of the crankshaft. This can be accomplished as the contact breaker unit retaining clip was released when the timing cover was removed. Note that the clip must be replaced with the threaded hole towards the inside of the engine.

6 In 1965 the vertical CB unit was replaced by a camshaft driven unit and a blanking plug was secured into the drive shaft aperture. This must be removed before the oil pump spindle, (which replaces the distributor drive shaft) can be driven upwards.

7 Next take off the sump cover and filter below the crankcase, and release the oil pump, for which there is just sufficient space to take it out through the timing case. The pump is retained by the three screws the other screws holding the pump assembly together. If the efficiency of the pump is in doubt it is advisable not to dismantle it but to obtain a new unit. Removal of the pump leaves the way clear for the insertion of a 3/8 in soft drift to the face of the distributor drive shaft (or oil pump spindle after 1964), which must be driven upwards until it is clear of the worm wheel on the engine mainshaft. It is important to note that on engines numbered from 9009, the distributor shaft bush is locked in position by a grub-screw which must be removed before the bush can be moved.

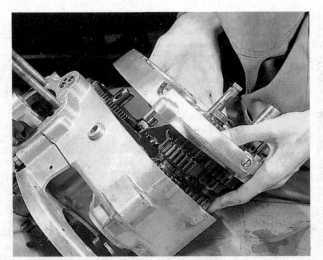

11.3 Removing inner cover and gear cluster

12.1a Withdraw the camshaft after making a note of the timing marks ...

12.1b ... and then slide out the tappets

12.3 A method of removing the crankshaft pinion

12.4 Taking off the oil pump

13 Splitting the crankcase halves

1 On the left hand side of the crankcase, remove the three bolts at the lower front of the case, the two nuts at the centre and the remaining two nuts below the cylinder barrel mouth.

2 Separate the crankcase halves by gently tapping them apart with a hideface mallet. On no account try to force the cases apart using a screwdriver as damage to the soft aluminum mating surfaces will occur resulting in oil leaks. By tapping the timing side of the crankshaft with a hide mallet the drive side of the crankcase should come away complete with flywhee assembly.

3 Support the drive side crankcase on wooden blocks and carefully tap out the flywheel assembly. On 350 and 344 models some force will be required because of the double bearing assembly. Note the number of shims, if any, fitted between the right hand side of the flywheel and main bearing on B25 models only.

14 Removing the main bearings

250, 350 and 441 cc engines:

The inner and outer races of the left side roller bearing are separated as the crankcase halves are split. The outer race can be driven out after the case has been warmed with hot water. The inner race, remaining on the crankshaft, can be pulled off using a suitable gear puller or BSA tool number 61-3778. The right side (timing side) ball bearing assembly can be driven out after warming the case.

Note: On earlier 250 and 350 models the timing side main bearing comprises a plain bush which can be driven out with an aluminum drift after heating the crankcase.

500 cc engines only:

1 To remove the two drive (left) side bearings, the bearing retaining ring attached to the crankcase by four countersunk screws - must first be taken off. Warm the crankcase and drive out the bearings, spacer, and abutment ring. The timing side bearing can be driven out after the right crankcase has been warmed.

2 Install new bearings in the same manner , after the cases have been warmed. Make sure that the abutment ring is correctly located. Apply a drop of thread locking compound to the four retaining ring screws upon installation.

15 Dismantling the big end and flywheel - B25 models

1 The connecting rod can be removed by simply unbolting the bearing cap. Loosen the nuts alternately, a turn at a time to prevent distortion. To facilitate reassembly, the connecting rod and cap have been marked with a centre punch. Note the direction in which the marks face.

2 Examine the bearing shells and crankpin carefully for signs of wear, scoring, and other damage. If it is necessary to regrind the crankshaft, bearings are available in 0.010, 0.020 and 0.030 in undersizes. It is very important that the radius at either end of the crankpin is machined to 0.070-0.080 in when regrinding. **Do not** attempt to refinish the bearing shells or file the bearing cap mating surfaces to reduce bearing clearances.

3 If the crankshaft is to be reground the flywheels must be removed. Loosen the four short flywheel retaining bolts (closest to the crankpin) first to avoid distortion. Remove the remaining four bolts and separate the flywheels. It would be a good idea at this time to clean the oil sludge trap, located in the right flywheel. Unscrew the plug and clean the passage with solvent and if available, compressed air.

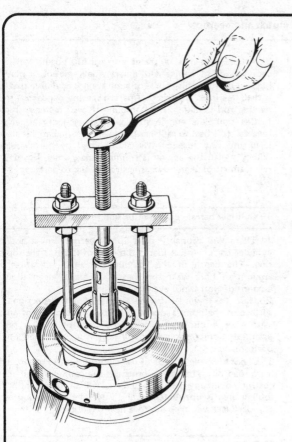

Fig. 1.6. Removing the crankshaft drive-side main bearing

13.1 Splitting the crankcase halves

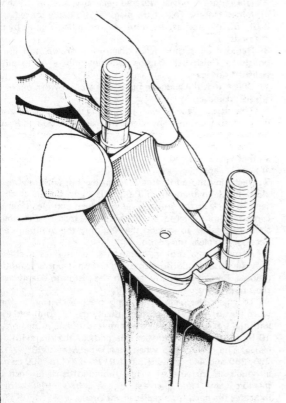

Fig. 1.7. Fitting a new big-end shell (B25 only)

15.1 B25 Flywheel assembly

16 Big end and flywheel assembly - 250 and 350 Star, B44 and B50 models

The above models are fitted with a roller bearing big end; the replacement of which necessitates the use of specialised equipment, ie a press for removal and installation of the flywheels and a special jig for truing them after assembly. Therefore, if it is decided that the big end requires replacement, a complete new flywheel assembly should be obtained from the nearest BSA stockist under a service exchange scheme.

1   The alloy connecting rod fitted to the BSA singles is very prone to cracking in the vicinity of the big end eye. The connecting rod should be cleaned and examined closely for the fault which usually occurs in the form of a hair line crack in the 12 o'clock position moving upwards quite vertically.

17 Main bearings and oil seals - examination

1   When the bearings have been pressed from their housings, wash them with a petrol/paraffin mix, to remove all traces of oil. If there is any play in the ball or roller bearings, or if they do not revolve smoothly, new replacements should be fitted. With regard to the phosphor bronze bushes, any signs of wear or roughness will immediately be evident. Replace them if there is any doubt about their condition.

2   Do not attempt to refit the old oil seals as oil leaks are bound to result. When fitting the left hand crankcase seal and the gearbox driveshaft seal push them in as far as possible by hand and then tap fully home using a wooden drift. They should face inwards, the "spring" side in.

3   Do not forget to check the camshaft bushes, (one in the right hand crankcase half and one in the gearbox inner cover) also check the kickstart needle bearing and the bush in the outer timing cover for excessive wear and replace if necessary.

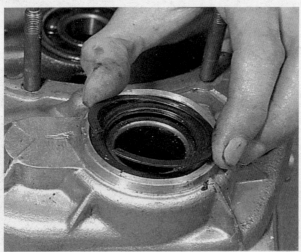

17.2 Fit new gearbox oil seal and insert circlip

18 Crankshaft assembly - examination and renovation

1   Wash the complete flywheel assembly with a petrol/paraffin mix to remove all surplus oil. Then hold the connecting rod at its highest point of travel (fully extended) and check whether there is any vertical play in the big end bearing by alternately pulling and pushing in the direction of travel. If the bearing is sound, there should be no play whatsoever.

2   Ignore any sideplay unless this appears to be excessive. A certain amount of play in this direction is necessary if the bottom end of the engine is to run freely.

3   Although it may be possible to run the engine for a further short period of service with a very small amount of play in the big end bearing, this course of action is not advisable particularly if shell bearings are fitted. Apart from the danger of the connecting rod breaking if the amount of wear increases rapidly, a further complete engine strip will be necessary to effect the renewal. It is best to replace the big end bearing at this stage, if it is in any way suspect. Wear is denoted by the characteristic 'knock' when the engine is running under load. For details on replacement of big end bearings refer back to paragraphs 15 and 16.

19 Cylinder barrel - examination and renovation

1   There will probably be a lip at the uppermost end of the cylinder barrel, which marks the limit of travel of the top piston ring. The depth of the lip will give some indication of the amount of bore wear that has taken place, even though the amount of wear is not evenly distributed.

2   Give the cylinder barrel a close visual inspection. If the surface of the bore is scored or grooved, indicative of an earlier seizure or a displaced circlip and gudgeon pin, a rebore is essential. Compression loss will have a very marked effect on performance.

3   Check that the outside of the cylinder barrel is clean and free from road dirt. Use a wire brush on the cooling fins if they are obstructed in any way. The engine will overheat badly if the cooling area is obstructed in any way. The application of matt cylinder black will help improve heat radiation.

20 Piston and piston rings - examination and renovation

1   Attention to the piston and piston rings can be overlooked if a rebore is necessary because new replacements will be fitted.

2   If a rebore is not considered necessary, the piston should be examined closely. Reject the piston if it is badly scored or if it is badly discoloured as the result of the exhaust gases by-passing the rings.

3   Remove all carbon from the piston crown and use metal polish to finish off. Carbon will not adhere so readily to a polished surface.

4   Check that the gudgeon pin bosses are not worn or the circlip grooves damaged.

5   The outside face of each piston ring should possess a smooth metallic surface and any signs of discolouration indicates gas leakage and means that the rings are in need of replacement.

6   The rings should also retain a certain amount of 'springiness' so that when released from the barrel, the ends of each ring are at least ¼" apart.

7   Each ring should be free in its groove but with minimum side clearance (a maximum of 0.002" is permissible). If the rings tend to stick in the grooves, remove them and clean out all the carbon from the groove and the inside face of the ring. Care should be taken to expand the ring only the minimum amount necessary to clear the piston.

8   A piece of a broken piston ring, ground as a chisel, will provide a useful tool for removing carbon deposits from the ring grooves. Be careful not to scratch the piston to deeply or widen the ring grooves.

9   To check the piston ring gaps, place each ring in the least worn part of the cylinder bore (usually the bottom) and locate it with the top of the piston to ensure it is square in the bore.

10 Measure the gap between the ends of the ring with a feeler gauge. The correct gap when new is between .009" - 013" for 250, 350 and 441 cc models and .017"-.025" for 500 cc models, although an increase of a few thousandths of an inch is permissible, any large increase to, say .030" (040" for B50) indicates the need for a replacement ring.

11 It is advisable to check the gap of a new ring before fitting, and if the gap is less than .007" the ends of the ring must be carefully filed to the correct limit.

12 Both compression rings on the B44 and B50 models are tapered on the outside face and their upper surface is marked "top" to ensure correct fitting. B40 and B25 models have a top compression ring of plain section and the second compression ring only is tapered this is also marked "top".

13 If these tapered rings are fitted upside down, oil consumption will become excessive.

Note: Earlier 250 Star models had two plain compression rings; the taper compression ring was introduced at engine No 2171.

21 Small-end - examination and renovation

1 If the con-rod is fitted with a small end bush check for wear by inserting the gudgeon pin through the bush. It should slide and rotate freely with no rocking movement. If play is present check the pin for smoothness and freedom from ridges, if satisfactory the bush has worn and must be replaced as follows.

2 The bush can be changed in one operation by pushing the old bush out and at the same time, pressing the new one in with service tool No 61-3653 (B44 and B50), No 61-3794 (B25/C25). The new bush must be correctly aligned with the oil hole and reamed to .7503-.7506 in (441 and 500 cc models) or .6890-.6894 in after pressing into the connecting rod.

3 Some later 250 cc models are not fitted with a replaceable bush and if excessive wear has occurred the complete conrod must be replaced.

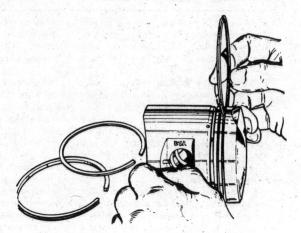

Fig. 1.8. Checking the piston ring grooves for wear

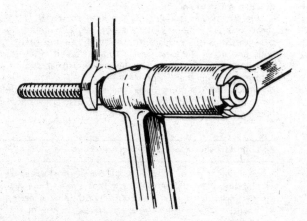

Fig. 1.9. Removing a small end bush (when fitted)

22 Cylinder head - examination and renovation

1 On 250 and 350 machine the cylinder head can be removed with the engine still in the frame, (see paragraph 7 operations 1 to 5). The 441 and 500 cc model do not have sufficient clearance between the cylinder head and the frame top tube to enable it to be lifted off and the engine will have to be removed from the frame. An alternative method however, is to extract the five central rocker box studs to allow the rocker box to be removed so providing the necessary clearance for cylinder head removal. Continual extraction of these studs however, will eventually impair the threads in the head and it is preferable to remove the complete engine.

2 Examine the rocker pads for wear; if it is excessive they will make correct valve clearances difficult to obtain and must therefore be replaced. Do not attempt to grind the pads smooth as they are case hardened.

3 If the rockers and spindles are dismantled take care to renew any damaged washers. On B44 and B50 rocker spindles, see that the rubber sealing rings are in good condition.

On the C25 models on and after engine No C25-2050 a redesigned rocker box assembly was fitted. The old type rockers will not fit the new type rocker box and vice-versa.

4 To remove the valve springs use BSA service tool No G1-3340 or a similar spring compressor, compress each spring until the split collets can be removed. The valve springs and top collars can now be lifted from the valve stems, cleaned in paraffin then placed on a numbered board to indicate their position in the cylinder head.

5 The springs may have settled through long use and they should therefore be checked in accordance with the dimensions given in the specification at the beginning of this Chapter.

6 Check the valves in their guides for excessive side-play. If excessive clearance exists, or if the valve stems are scored or have a carbon build-up on them, the valves and guides should be replaced. Valve guides can be driven out using an aluminum drift, after the cylinder head has been heated in an oven or immersed in hot water. Install the new guides while the head is still warm. Note that the exhaust valve guide is counterbored at its lower end.

7 If the valve faces are badly fitted or burned they must be replaced. The exhaust valve is usually in the worst condition because of the hot gases that flow over the sealing face. Sometimes, if the valves have been reground quite a few times the valve seats become recessed or "pocketed". To obtain the best from an engine the area of metal in the cylinder head should be cut back from around the valve seat as shown in the accompanying illustration. This is a highly skilled operation and should only be entrusted to a suitably qualified engineer.

8 Remove the carbon from the combustion chamber using an aluminum scraper. Be careful not to scratch the soft metal of the cylinder head, especially the valve seats. Clean the head in paraffin after the carbon has been removed.
CAUTION: Do not use a caustic soda solution to clean aluminium parts.

9 Whether the original valves or new valves are to be used, they should be lapped into their seats. Put three small dabs of grinding compound around the valve face and, using the valve grinding tool as shown, rotate the valve back and forth. Do not use pressure. Lift and rotate the valve to a new position every few seconds and continue grinding until a smooth, even finish is obtained on the valve and valve seat.

10 Before assembling the valves, all traces of grinding compound and grit must be removed from the head and valves. Coat the valve stems with fresh oil before installing. Assemble the valve springs and retainers as removed. A small amount of grease will keep the split collar in place as the spring compressor is removed. Make sure the collar is correctly seated by tapping the valve stem with a small, soft hammer.

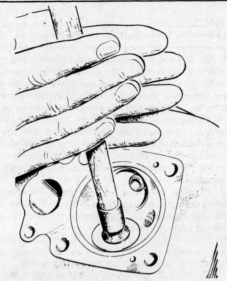

Fig. 1.10. Regrinding the valves

B25.C25 Only   **Other Models**

Fig. 1.11. Check the rocker pads for wear

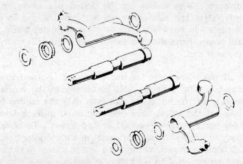

Fig. 1.12. Rocker assembly (all models except B25/C25)

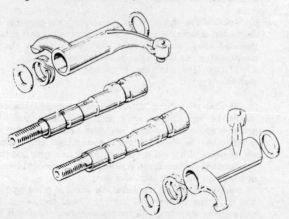

Fig. 1.13. Valve rocker assembly (B25/C25). (The spring washers were not fitted to models made on and after engine No. C25-2050)

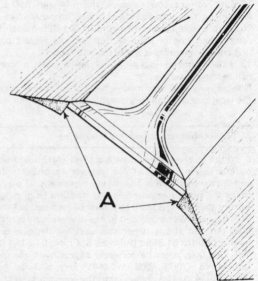

Fig. 1.14. 'Pocketed' valve, showing the metal that should be removed (shaded area A)

23 Oil pump - examination and renovation

1   Unless it is suspected that the oil pump internal gears may be damaged or worn through negligence in changing the oil at the specified periods, the pump should not be dismantled. However, if an inspection of the internal components is required proceed as follows.

2   Unscrew the four screw from the base of the pump and take off the base plate and top drive housing from the pump body.

3   The driving spindle and driving worm gear are secured to the top cover with one nut and spring washer. Before removing the worm gear, make careful note of the way in which it is fitted to assist in rebuilding. Note also the position of thrust washers below top gears (B44 and B50 models only).

4   Wash all the parts thoroughly in paraffin allow to dry before examining. Look for foreign matter jammed in the gear teeth and deep score marks in the pump body. These will be evident if the oil changing has been neglected. Slight marks can be ignored, but any metal embedded in the gear teeth must be removed.

5   The most likely point of wear will be found on the driving gear teeth; if these are worn to the extent that the sharp edges have gone then they must be renewed.

6   When rebuilding the pump ensure that all the parts are clean and free from dirt or grit. Insert the driving spindle (with fixed gear) into pump top cover, fit the worm drive and secure in position with nut and spring washer.

7   Fit the driven spindle and gear into the cover and replace thrust washers. The oil pump used on B25/C25 models does not have thrust washers. Place the assembly on top of the pump body and insert the lower gears. Apply clean oil to the gears and refit the base plate. Check that the spindle and gears rotate easily before tightening the four fixing screws.

8   Check that the pump body face is not distorted by holding it up to the light and resting the edge of a steel rule across the housing face.

24 Camshaft and tappets - examination

1   Examine the foot of each tappet for signs of excessive wear or chipping. The bottom of each foot should have a smooth mirror finish and if grazed or blemished they must be replaced.

2   Examine the camshaft lobes in the same way as described above and replace if necessary. Also check the timing gears for broken or badly worn teeth.

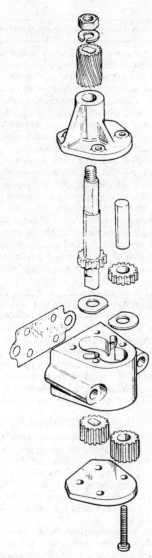

Fig. 1.15. Exploded view of oil pump (later models)

Fig. 1.16. Exploded view of oil pump (engines up to 1966)

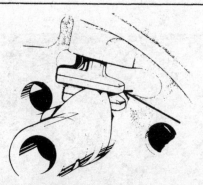

Fig. 1.17. Correct position of cam followers

25 Push rods - examination

Examine the push rod cup and ball ends for excessive wear or chipping and check that the rods are straight by rolling them on a piece of glass or any true, flat surface. If any faults are evident the rod(s) must be replaced.

26 Crankcase - examination and renovation

1   Inspect the crankcases for cracks or any other signs of damage. If a crack is found, specialist treatment will be required to effect a satisfactory repair.

2   Clean off the jointing faces, using a rag soaked in methylated spirit to remove old gasket cement. Do not use a scraper because the jointing surfaces are damaged very easily. Check also the bearing housings, to make sure they are not damaged. The entry to the housings should be free from burrs or lips.

3   Do not forget to check also the gearbox inner and outer covers and the primary drive cover. Good jointing surfaces are essential especially in the case of the gearbox inner and outer covers that have no intermediate gasket.

27 Oil pressure valve - examination

1   To maintain the engine oil system at a constant pressure a relief valve is on the front right hand side of the crankcase. If oil pressure becomes excessive the valve opens and allows oil to pass back into the crankcase.

2   Remove the valve by unscrewing the large nut and draw out the spring and ball. If either components are corroded or the spring weak replace with new items. On no account stretch the spring or attempt to alter the valve setting.

3   On 1970 models an improved oil pressure release valve has been fitted in place of the ball and spring type. It consists of a hexagon body with a gauze covering the end which screws into the crankcase. Into this is fitted a piston with a spring behind it. This in turn is secured by a domed cap which screws onto the end. Examine the valve assembly as for earlier types and replace components where necessary.

28 Oil sump filter - examination

1   Remove the four nuts holding the oil sump plate onto the bottom of crankcase and withdraw filter. Note that on C15 models it is necessary to slacken the engine mounting bolts to allow sufficient clearance for sump plate removal. Clean thoroughly in paraffin and dry. While the sump cover is off check the operation of the scavenge non-return valve located inside the sump by pushing the ball off its seat with a piece of wire and checking that it falls back onto its seat when the wire is withdrawn.

2   If the ball sticks, clean the passageway with petrol until the valve functions correctly. Replace the filter and sump cover using new gaskets.

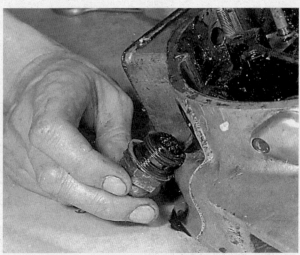

27.2 Remove the oil pressure valve for cleaning

27.3 Oil pressure valve dismantled

28.2 Replacing sump filter after cleaning

29 Gearbox components - examination and renovation

1   Examine carefully the gearbox components for signs of wear or damage such as chipped or broken teeth on the gear pinions and kickstarter quadrant, rounded dogs on the ends of the gear pinions, bent selector forks, weakened or damaged springs and worn splines. If there is any doubt about the condition of a part, it is preferable to play safe and replace the part at this stage. Remember that if a suspect part should fail later, it will be necessary to completely strip the engine/gear unit yet again.

2   It is advisable to replace the kickstarter return spring irrespective of whether it seems to be in good condition. This spring is in constant use, yet if it has to be replaced at a later date, a certain amount of dismantling is necessary in order to gain access. It is cheap and easy to replace at this stage.

3   Do not forget to examine also the kickstarter ratchet assembly. Examination will show whether the ratchet teeth have worn, causing the kickstarter to slip or whether the outer teeth are damaged, causing the kickstarter quadrant to jam. Note that the leading tooth of the quadrant is relieved, to help offset the tendency to jam during the initial engagement.

30 Clutch assembly - examination and renovation

1   The four driving plates have segments of special friction material which are securely bonded in the metal. These segments should all be complete, unbroken and not displaced. If any segments are damaged the complete plate should be exchanged for a new one. The overall thickness of each segment should be measured and if the extent of wear is more than .030" (.75 mm), the plates should be replaced. Standard thickness is .167" (4.242 mm).

2   The tags n the outer edge of the plates should be a reasonable sliding fit in the chainwheel slots and should not be "hammered" up. If there are burrs on the tags they must be carefully dressed out with a fine cut file. If the tags are badly damaged the plate must be replaced.

3   The plain driven plates should be free from score marks and prefectly flat. To check the latter, lay the plate on a piece of plate glass; if it can be rocked from side to side, it is buckled and should be replaced.

4   To inspect the cush drive rubbers within the clutch centre, take out the four countersunk head screws adjacent to the clutch spring housings and prise off the retaining plate.

5   If rubbers are quite firm and sound they should not be disturbed unless wear or damage is suspected.

6   When refitting the clutch rubbers it may be found necessary to use a lubricant, in which case a liquid soap is recommended.

7   Examine the slots in the clutch chainwheel for wear; if they are corrugated or the teeth are hooked and thin, the chainwheel should be replaced.

8   Check the chainwheel roller bearing for up and down movement. Slight play is permissible but if excessive, the bearings should be renewed.

9   The clutch operating rod should be checked for distortion by rolling it on a piece of plate glass. If bent it should be replaced or, straightened by means of a surface block and a very light hammer. The correct length for the rod is 9 inches and if it has worn below this length it should be replaced.

10 Also check the clutch springs - these should have a free length of approximately 1-6 inches and if the uncompressed length is well below this figure they should be replaced.

31 Primary chain - examination

1   The primary chain on all models is the Duplex endless type and because it is fully enclosed and running in an oil bath with provisions for adjustment, it should give long and trouble free service. However, check for any broken or cracked rollers and excessive wear in the links.

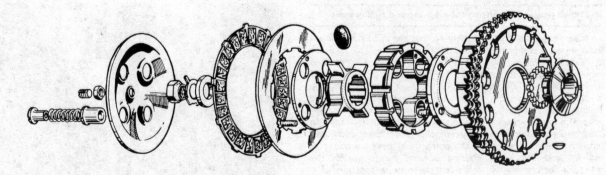

Fig. 1.18. Exploded view of clutch

2 An early indication that the primary chain is being starved of oil is the appearance at the joints of a reddish brown deposit, and this should be taken as a warning that there is something amiss with lubrication. This could be caused by a leaking gasket or the fact that the oil seal behind the clutch is faulty; in both cases replacement parts should be fitted.

3 If any doubts exist regarding the integrity of the chain - replace it. A broken primary drive chain is not only inconvenient - but can also be dangerous and expensive.

32 Reassembly - general

1 Before the engine, clutch and gearbox components are re-assembled, they must be cleaned thoroughly so that all traces of old oil, sludge, dirt and gaskets are removed. Wipe each part clean with a dry, lint-free rag to make sure that there is nothing to block the internal oilways of the engine.

2 Lay out all the spanners and other tools likely to be required so that they are close at hand during the reassembly sequence. Make sure the new gaskets and oil seals are available - there is nothing more infuriating than having to stop in the middle of a reassembly sequence because a gasket or some other vital component has been overlooked.

3 Make sure the reassembly area is clean and unobstructed and that an oil can with clean engine oil is available so that the parts can be lubricated before they are reassembled. Refer back to the torque wrench settings and clearance data where necessary. Never guess or take a chance when this data is available.

4 Do not rush the reassembly operation or follow the instructions out of sequence. Above all, do not use excess force when parts will not fit together correctly. There is invariably good reason why they will not fit, often because the wrong method of assembly has been used.

33 Engine reassembly - fitting bearings to crankcase

1 Before fitting the crankcase bearings, make sure that the bearing surfaces are scrupulously clean and that there are no burrs or lips on the entry to the housings. Press or drive the bearings into the cases, using a mandrel and hammer, after first making sure that they are lined up squarely. Warming the crankcases will help when a bearing is a particular tight fit.

2 When the bearings have been driven home, lightly oil them and make sure they revolve freely. This is particularly important in the case of the main bearings. There is one (two on the B50) on the drive side and one bearing on the timing side left hand and right hand crankcases respectively) each of the journal ball type.

3 Using a soft mandrel, drive the oil seals into their resepctive locations. Do not use more force than is necessary because the seals will damage very easily.

4 Do not omit the circlips that retain the bearings in position.

34 Engine reassembly - fitting the crankshaft assembly

1 On B25 engines, crankshaft end float must be restricted to .002"-.005". This is controlled by shims fitted between the crank web and the inner face of the right bearing. Shims are available in thicknesses of .003" (40-0064), .005" (40-0065), .010" (40-0066), and .015" (40-0069).

2 In the case of B50 engines ensure that the bearing abutment ring on the drive side is correctly located, and apply a drop of "Loctite" to the threads of each of the four screws securing the bearing retaining plate.

3 Place the crankshaft assembly into the timing side crankcase. This operation will be simplified if the case is supported on wooden blocks so that the crankshaft is clear of the bench top.

4 Apply a thin coating of gasket cement to the joint faces of each crankcase half and fit the drive side gear.

5 Replace the three bolts at the front of the case and the four nuts (two at the base of the cylinder and two in the primary case). Tighten bolts and nuts evenly, to avoid distorting the joint faces.

6 Check that the crankshaft assembly rotates quite freely. If it does not, then the alignment may be incorrect and the cause of the trouble must be rectified.

34.3 Refitting the crankshaft assembly

35 Engine reassembly - fitting the oil pump and timing gears

1   Invert the crankcase assembly so that the right hand side is facing upward and proceed as follows:-

2   Slide the distance piece onto the driveshaft ensuring that the chamfered side faces outwards. This distance piece is not fitted to earlier 250 and 350 models.

3   Next push the pump drive gear on making sure that the keyway faces outwards. Smear a thin coating of grease on both sides of a **new** pump gasket, make sure that the mating faces are clean and the oilways unobstructed, then slide gasket and pump onto the studs. Ensure that the pump drive gears are correctly engaged and the pump housing is lying evenly against the crankcase before gently tightening the nuts in rotation to a torque of 5 - 7 lbs ft.

Note: New self locking nuts should be used.

4   Replace the crankshaft timing pinion with the timing mark(s) facing outward and slide the tappets into their housings making sure that the thinner end of each foot faces towards the front of the engine. On the earlier 250 Star engine the tappet must be replaced with the oil holes in the stem towards the rear of the engine.

5   Fit the camshaft and timing gear paying careful attention to the timing marks. On some machines there is a single dash on each timing gear and these should be aligned as shown in the accompanying illustration. If however, there is a dash and a V the dash must be ignored and the V on the camshaft gear aligned with the dash on the crankshaft gear.

6   Having correctly set the timing gears fit the tab washer and screw on crankshaft nut - to tighten place a suitable rod through the small end bearing to prevent rotation of crankshaft assembly.

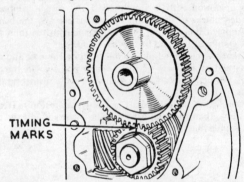

TIMING MARKS

Fig. 1.19. Engine timing marks

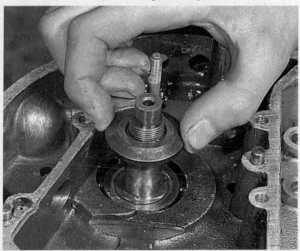

35.2a Do not forget to fit the thrust washer ...

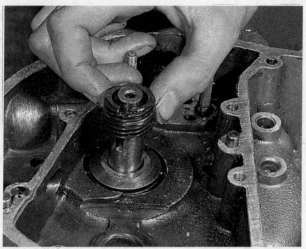

35.2b ... before installing oil pump drive gear

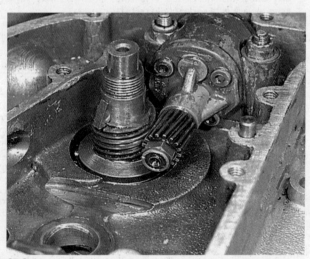

35.3 Correct installation of oil pump

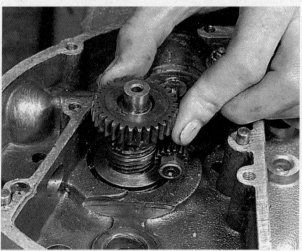

35.4a Fit the crankshaft pinion ...

35.4b ... and then the tappets

35.5 Check timing marks carefully before installing camshaft gear

35.6 Finally fit pinion tab washer and nut

36 Gearbox reassembly - fitting gear cluster to inner cover

1 It is essential that all oil seals have been renewed and bearings and bushes have been replaced where necessary.

2 Place the inner cover, inside face up, on the bench and replace the cam plate, correct way round, in the cover slot (see adjacent photo for guidance), insert the pivot pin and secure with the split pin.

3 Insert the mainshaft fitted with its low gear and third gear, into the cover bearing, replace the kickstart ratchet assembly and secure with the fixing nut. It will be necessary to hold the mainshaft in a vice, using soft metal clamps, to tighten the nut fully.

4 Holding the cover face down, place the layshaft low gear with its shim and sliding gear (third) in position on the cover. Fit it's selector fork, the roller being located in the lower camplate track.

5 Next fit the mainshaft sliding gear (second) with the appropriate spacers. Replace its selector fork and locate the fork roller in the upper cam track. Insert the spindle through the selector fork bosses and locate in the cover.

6 The layshaft, with its remaining two gears (fixed high gear and second gear) can now be passed through the gears on the cover, into kickstart boss aperture.

7 Fit the gearchange return spring to the quadrant and replace the assembly in the cover, locating the spring loop over the pivot bolt. It will be necessary, whilst carrying out this operation, to press in the plungers with a knife before finally engaging the plungers with the cam plate slots as the quadrant is pressed home.

8 A thrust washer is fitted to the drive side end of the layshaft where there should be just perceptible end float. The mainshaft, being locked to the inner cover, does not need checking for end float but excessive movement between the gears and the ends of the splines must be corrected by fitting the appropriate spacers. See exploded view of gear cluster for position of each spacer.

9 Fit the kickstart quadrant shaft through the front of the cover ensuring the internal needle roller bearing slides over the end of the layshaft. Tension the kickstart spring and hook the retaining eye over the stud.

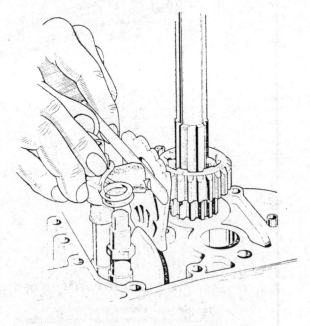

Fig. 1.20. Refitting the gearchange quadrant spring

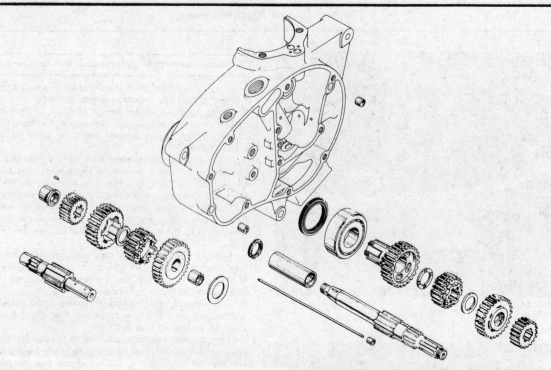

Fig. 1.21. Exploded view of earlier 250 and 350 Star gearbox cluster

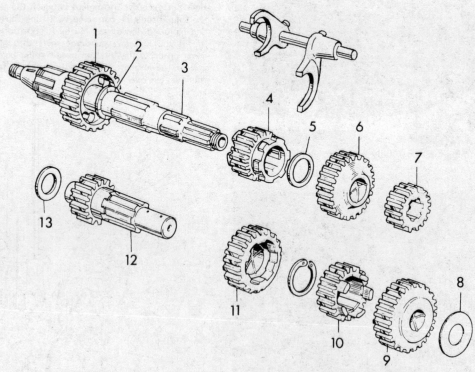

FIG. 1.22. EXPLODED VIEW OF GEAR CLUSTER FITTED TO ALL LATER MODELS

1 Mainshaft sleeve pinion (top gear)
2 Mainshaft sleeve pinion thrust washer
3 Mainshaft
4 Mainshaft sliding gear (second gear)
5 Mainshaft third gear thrust washer
6 Mainshaft third gear
7 Mainshaft bottom gear

8 Layshaft bottom gear shim
9 Layshaft bottom gear
10 Layshaft sliding gear (third gear)
11 Layshaft second gear
12 Layshaft
13 Layshaft thrust washer

36.2a Cam plate fitted to inner cover

36.2b Inserting cam plate retaining pin

36.3a Replace the kickstart dog ...

36.3b ... followed by the driving gear, spring and sleeve

36.3c Finally tighten nut

36.5 Mainshaft gears fitted to cover

36.7 Correct position of selector forks in cam plate and quadrant spring

37.1 Fit new oilway 'O' ring before installing inner cover

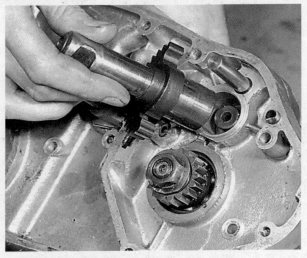

36.9 Inserting the kickstart quadrant

37 Gearbox reassembly - fitting inner cover and gear cluster to crankcase

1 First check that the oilway 'O' ring is in position as shown on accompanying photograph. Carry out a thorough re-check to make sure that all shims are in place where necessary and that the gears and selector mechanism have been assembled correctly. It is far better to discover a fault at this stage than when you take the bike out for its first run.
2 If you are satisfied that everything is correct, smear a coating of jointing compound around the crankcase face and carefully slide the gear assembly into the gearbox housing. If it sticks before the two faces mate together, **do not use force** instead apply a steady pressure with one hand and guide the mainshaft through the gearbox drive aperture with the other hand.
3 When the inner cover is correctly fitted replace all the retaining screws. Temporarily slide the gearchange lever onto the gear selector shaft and check that the four gears and neutral can be selected - if satisfactory remove the gearchange lever.

38 Engine reassembly - fitting the primary drive

1 Refit the gearbox drive sprocket and lock washer and screw on the retaining nut. To tighten the nut wedging the gearbox sprocket with a piece of wood or by wrapping a chain around it.
2 Fit a new seal into the primary drive dust cover and replace cover using a new gasket - tighten the six retaining screws evenly.
3 Refit the engine crankshaft distance piece and replace the clutch sleeve onto the gearbox mainshaft, checking that the woodruff key is in position. Note that the distance piece, where fitted, has a chamfered edge on one side; this should be installed in the direction noted during removal.
4 Attach the primary chain around the clutch chainwheel and engine sprocket, making sure that the protruding boss on the engine sprocket faces outwards. Check that the woodruff key on the engine shaft is correctly positioned and replace the engine sprocket and clutch chainwheel simultaneously onto their respective shafts.
5 Centralise the clutch housing on the clutch sleeve, smear plenty of grease around the sleeve and insert the 25 roller bearings.
6 Replace the clutch hub onto the splines of the sleeve; fit the thick washer with the recess outwards onto the gearbox mainshaft followed by the tab washer and nut. Jam a piece of wood between the primary chain and clutch chainwheel and tighten the nut. Finally bend up the tab washer.
7 Replace the clutch plates starting with one plain then one friction plate and so on alternately, there being five plain and four friction plates. Insert the clutch push rod into the hollow mainshaft.
8 Place the pressure plate in position and fit the four spring cups and springs, which should be of equal length. If in any doubt about the condition of the springs, replace them since they are quite inexpensive.
9 Fit the special spring nuts, (4 off) and tighten down evenly until the first coil of each spring is just proud of its cup. If the springs are tightened excessively the clutch will drag causing difficulty if changing gear, if the springs are not done up sufficiently, clutch slip will result. On B50 engines the spring retaining nut must be wirelocked.

38.1 After tightening gearbox sprocket retaining nut, bend up lockwasher

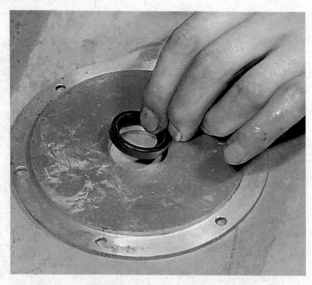

38.2 Fitting a new seal in primary drive dust cover

38.3a Installing engine drive shaft spacer

38.3b Clutch sleeve in position

38.4 Fitting the primary drive chain

38.5 Clutch roller bearings in position

38.6 Fit the clutch hub — followed by the washer and nut

38.7 Install the clutch plates in the correct sequence

38.9 Tighten clutch spring nuts evenly

39 Engine reassembly - fitting the alternator

1   Slide the rotor onto the keyed engine shaft making sure that the 'Lucas' trademark faces outwards - fit the tabwasher and nut and after tightening the latter bend up the tabwasher.

2   Replace the primary chain tensioner onto the two lower stator studs ensuring the spacer on the rear stud is in position.

3   Fit the stator on to its studs with the lead on the inside at the front, and secure with the self-locking nuts. It is important that the air gap between the rotor and the stator pole pieces is equal all round. The gap may be checked with a .008'' feeler gauge. Any variation may be corrected by slackening the stator fixing nuts sufficiently to allow the stator to be tapped into the required position with a hide mallet.

4   Having refitted the stator, checked the air gap and tightened the three fixing nuts, slacken the rearmost nut in order to adjust the primary chain. Adjust the chain tensioner to permit ¼'' (6 mm) free play on the top run of the chain midway between the sprockets, and tighten the stator nut firmly.

5   The primary drive cover can now be replaced using a new gasket lightly smeared with grease. When replacing the retaining screws ensure they are replaced in the correct positions depending on their length.

40 Engine reassembly - refitting the piston and cylinder barrel

1   Refit the piston complete with piston rings to the connecting rod. If the advice given for dismantling the engine has been followed, the inside of the piston skirt will have been marked so that the piston is fitted in the same position.

2   If the gudgeon pin is a tight fit, it may be necessary to warm the piston in order to expand the gudgeon pin bosses. Holding the piston under a stream of hot water will usually suffice, provided the water is quickly wiped off before reassembly commences. Push the pin home and replace both circlips, which should have been renewed. It is false economy to reuse the originals, which may become displaced and cause serious engine damage.

3   It is wise to pad the mouth of the crankcase with rag whilst the circlips are being fitted, to prevent them from falling in. If a misplaced circlip drops in, it may be necessary to dismantle the complete engine in order to reclaim it.

4   Check that both circlips are located in their grooves.

5   Before the cylinder barrel is replaced, check that the two nuts at the top left hand side of the crankcase are loose. Oil the piston and the inside of the cylinder barrel.

6   Fit a new cylinder base gasket and place two lengths of 1'' square wood on either side of the connecting rod, rotate the crankshaft until the base of the piston is resting firmly on the wood.

7   Space the three piston ring gaps approximately 120 degrees apart to ensure minimum gas leakage and slide the cylinder barrel over the four studs carefully feeding the piston in by compressing one ring at a time to ease the insertion of the rings the bottom of the barrel has a slight chamfer.

8   If available a piston ring clamp can be used but make sure that the clamp is not overtightened thus preventing it from sliding off the piston. Once the rings are inside the barrel unfasten the clamp, remove the wooden blocks and push the barrel down until it seats on the new base gasket.

9   Insert the two push rods down the barrel aperture, on to their respective tappets, the outer one operating the inlet valve (see accompanying illustration).

10 On B25 and B44 models only, the top of the exhaust push rod is painted red for identification purposes and is very slightly shorter than the inlet rod. The push rods of B50 engines are identical.

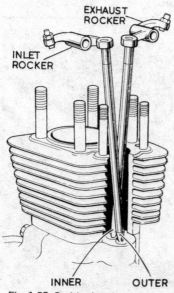

EXHAUST ROCKER

INLET ROCKER

INNER OUTER

Fig. 1.23. Positioning the pushrods

39.3 Pushing the alternator stator onto the three studs

39.1 Fitting the alternator rotor

39.5 Replacing the primary drive cover, note the oil level and drain screws

39.2 Chain tensioner in position

40.1 A piece of cloth placed as shown will prevent circlips etc., from dropping into the crankcase when refitting the piston

40.7 Fit the cylinder barrel using a new base gasket

41.1b ... before fitting the cylinder head

41 Engine reassembly - fitting the cylinder head and rocker box

1 Ensure the cylinder head face is clean; fit a new head gasket in position and slide the head over the studs taking care not to foul the pushrods. Tighten the six cylinder head nuts firmly and evenly.
Note: - (B44 only). To overcome any possible gas leakage at the cylinder head joint face, a gasket having a raised lip can be fitted in preference to the plain gasket. When fitting the new type gasket however, it is essential that the raised lip is uppermost.
2 Before fitting the rocker box the piston should be set at top dead centre (TDC) on the compression stroke to avoid any strain on the rocker box casting due to valve spring pressure.
3 Remove the rocker caps from the rocker box and fit a new gasket onto the top of the cylinder head. Slide the box over the studs; fit the pushrod ends into their appropriate rocker arms and push the box down onto the gasket. Double check that the pushrods are in their correct positions before tightening the rocker box nuts.
Note: If the engine is fitted with a valve lifter, do not forget to

41.2 Correct position of rockers (B25/C25 only)

refit the associated bracket.

42 Engine reassembly - checking the valve clearances - B50, B44, 250 and 350 cc Star models

1 The clearances between the top of each valve stem and the rocker adjusting pin, must be set when the engine is quite cold.
2 Remove the rocker caps and take out the spark plug and set the gearbox to neutral, to enable the engine to be rotated easily by hand.
3 Set the piston at top dead centre on the compression stroke (both valves closed) and using a feeler gauge, check that the fully open gaps for the inlet and exhaust valves are as follows:-
.008'' (inlet) and .010'' (exhaust)
4 If the clearance requires adjusting slacken the locknut and adjust the pin until the correct size feeler gauge will just slide between the valve stem and pin as shown in adjacent illustration.
5 Holding the pin in its new position, retighten the locknut. Check the clearance again to make sure that the setting has not altered whilst tightening the locknut.

41.1a Always use a new head gasket ...

B25 and BC25 models only:

6 Set the engine up as described in the preceding paragraphs. The B25/C25 are fitted with eccentric rocker spindles and valve clearance adjustment is carried out as follows.

7 Remove the cover plate on the right hand side of the rocker box, loosen the rocker spindle nut and turn the spindle inwards to obtain the correct valve clearance. It is most important that the spindle flats are in the vertical position before commencing and that they remain in the shaded sections, indicated in the accompanying illustration throughout adjustment.

8 During adjustment it may be found that one of the rocker spindles has reached the end of its thread and can no longer be turned forward, in which case the spindle must be turned back through 360 degrees before commencing adjustment.

9 When satisfied the correct clearances have been obtained replace the rocker cover using a new gasket and also the two inspection caps.

43.1 Inserting the clutch operating rod

FLATS ON SPINDLES IN NEUTRAL POSITION

Fig. 1.24. Valve rocker adjustment (B25/C25 only)

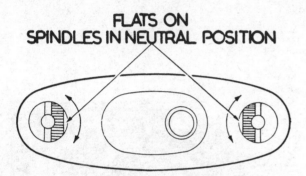

43.2 Check the release rack and ball — use grease to hold the ball in place

42.4 Adjusting the tappets (B25/C25 only)

43 Engine reassembly - fitting the outer timing cover

1 Lightly oil the clutch operating rod and insert it, with the bell-end outwards, through the centre of the gearbox mainshaft.

2 Ensure that the clutch release rack is correctly located inside the timing cover and place the ball bearing inside the rack using plenty of grease to hold it there.

3 Smear jointing compound onto the crankcase face and fit the cover making sure that the ignition lead grommet fits into the cutaway provided. Insert the retaining screws into their correct positions and tighten evenly.

43.3 Fitting the timing cover

44 Engine reassembly - ignition timing

B25, C25, B44 and B50 models:

1 Fit the automatic advance unit into the internal taper inside the end of the camshaft but **do not** force it fully home or insert the retaining bolt. Fit the contact breaker assembly into the timing cover but do not tighten the two screws.

2 Remove the small inspection cover at the front of the primary chaincase to expose the timing pointer and index mark. On engines with two index marks on the alternator rotor, use the mark identified by the number "2". On engines with two pointers, use the one appropriate to your engine as shown in the accompanying illustration.

3 Remove the rocker box inspection caps. Rotate the engine until the position is close to TDC of its compression stroke (both valves closed, clearance at both rocker arms). Turn the engine back a few degrees so that the timing mark goes past the pointer and then forward again in the normal direction of rotation until the pointer and mark are aligned. This will have taken up any backlash in the contact breaker drive and positioned the engine exactly where the spark occurs under full advance conditions.

4 Connect a test light using a motorcycle battery as shown. The exact time at which the points will be indicated by the light going out.

5 Revolve the cam using a pair of slim-nosed pliers until the fibre heel is on the peak of the cam, loosen the fixed contact screw and move the contact accordingly to give the correct gap of .015 in (.381 mm). Tighten the contact screw and recheck the setting.

6 Twist the contact breaker base plate as far as it will go in a clockwise direction and retain in this position by lightly doing up the pillar bolts.

7 Next, using the slim-nosed pliers, hold the auto-advance unit into its taper, at the same time rotating the unit anti-clockwise until the points are just opening and the light goes out. Be very careful not to mark the surface of the cam with the pliers.

8 Fix the auto-advance unit with the centre bolt and tighten. The timing is now set in the fully retarded position, and it will be necessary to set the timing fully advanced.

9 Slacken the pillar bolts and turn the contact breaker base plate anti-clockwise to its limit so that the timing light is again lit. Now turn the auto-advance unit cam against the spring-loaded weights to the fully advanced position using a screwdriver or similar tool applied in the slot of the cam, and rotate the contact breaker base plate clockwise until the points have just opened and the light is extinguished. The contact breaker base plate is now in its correct position and the pillar bolts can be tightened.

10 Finally recheck the ignition setting by turning the cam against the spring weights to the advance position ensuring the timing light goes out when the cam reaches the fully advanced position. If necessary fine adjustment can be made by slacking the two screws that retain the contact breaker assembly onto the base plate and turning the small eccentric screw at the bottom of the plate. Re-tighten retaining screw when final adjustments have been made.

Note: The fine adjustment facility is only fitted to models manufactured after 1968.

Earlier 250 and 350 Star models:

1 On earlier 250 and 350 cc Star model not fitted with the timing inspection cap on the primary chaincase the following method of ignition timing is recommended.

2 Obtain a length of 1/8" aluminum wire or a knitting needle of sufficient length not to disappear completely into the cylinder with the piston at BDC. Insert the wire through the plug hole, engage top gear and rotate the engine using the kickstart until the piston is at TDC on the compression stroke, (both valves closed).

3 Mark a thin line on the wire level with a reference point on the cylinder head, withdraw the wire and scribe another line

1/16" above the first, (0.007" on engine No 41600 and after).

4 Place the wire back in the cylinder and rotate the engine backwards by turning the gearbox driving sprocket, until the piston has fallen approximately one inch, then rotate the engine forward until the highest line on the wire is level with the previously marked reference point on the cylinder head. Keeping the piston at this position; remove the cap from the contact breaker unit behind the cylinder barrel and check that the points are just opening - this can be checked with a light bulb and battery as described previously.

5 If the points are closed, slacken the rearmost of the two screws at the top of the timing cover and rotate the contact breaker body until the points are just opening. Retighten the retaining screw and recheck the position of the wire in the cylinder barrel in relation to the opening of the points, Finally replace C.B. cover using a new gasket.

44.1a Fit advance and retard mechanism but do not fit the centre bolt yet

44.1b Replace contact breaker assembly and fit LT ignition lead

44.5 Check contact breaker gap before and after timing ignition

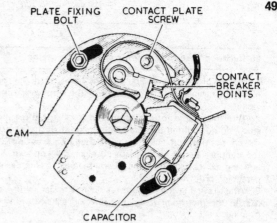

PLATE FIXING BOLT CONTACT PLATE SCREW

CONTACT BREAKER POINTS

CAM

CAPACITOR

Fig. 1.26. Contact breaker assembly (earlier models)

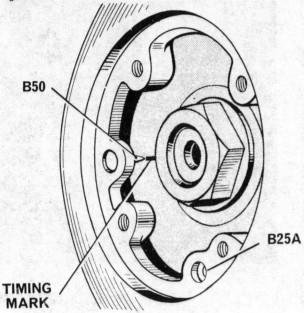

B50

TIMING MARK

B25A

Fig. 1.25. Ignition timing pointer positions (later models)

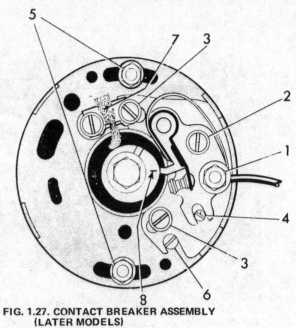

FIG. 1.27. CONTACT BREAKER ASSEMBLY (LATER MODELS)

1 Contact breaker lead securing nut
2 Contact adjustment screw
3 Baseplate securing screw
4 Eccentric gap adjustment screw
5 Breaker plate assembly securing screws
6 Eccentric base plate adjuster screw
7 Felt oiling wick
8 Breaker cam high point

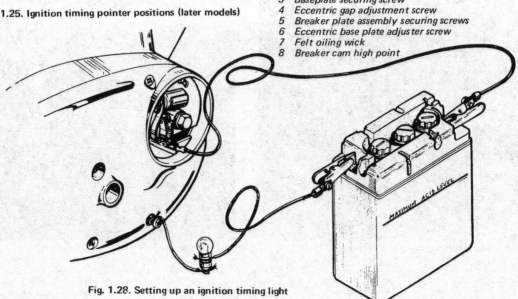

Fig. 1.28. Setting up an ignition timing light

45 Refitting the engine/gearbox unit into the frame

1   Before placing the engine into the frame, refit the engine oil feed/return pipes onto the bottom of the crankcase using a new gasket.

2   Lift the engine into position and insert the three mounting studs, do not tighten the retaining nuts until all three studs are in the correct position. Refit the spark plug to prevent dirt entering the cylinder barrel and replace all electrical connectors, cables and engine accessories in the reverse manner to that described in paragraph 5 in this Chapter.

3   When refitting the carburettor do not forget to fit a new 'O' ring in the mounting flange to avoid air leaks resulting in poor running.

46 Clutch adjustment

1   After refitting the clutch control cable to the operating lever on top of the gearbox, adjust the clutch using the following procedure.

2   Remove the small inspection cap at the rear end of the primary chaincase and slacken the locknut in the centre of the clutch pressure plate. Adjust the clutch by means of the centre screw; there should be a small amount of free-play on the handle bar control lever when the clutch is adjusted correctly, but not enough movement to cause clutch drag. After final adjustment tighten the locknut and replace the inspection cap.

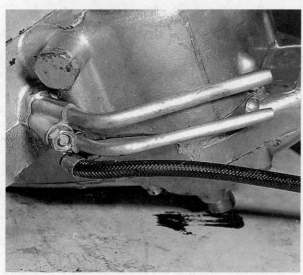

45.1 Correct location of engine oil return and feed pipes

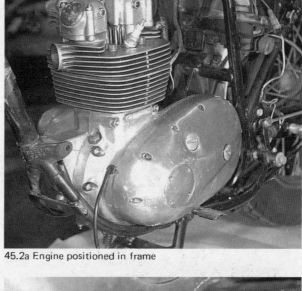

45.2a Engine positioned in frame

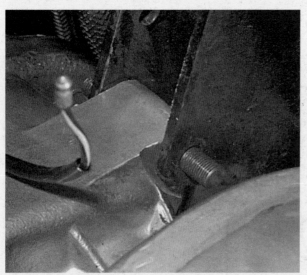

45.2b Insert the rear engine mounting stud first ...

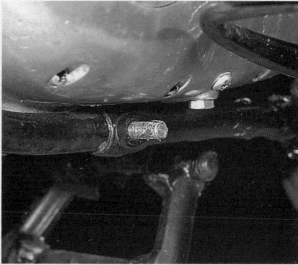

45.2c ... followed by the long centre stud ...

45.2d ... and finally the front stud and then tighten up all three

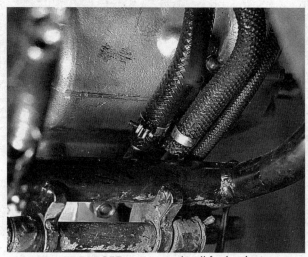

45.2e DO NOT FORGET to connect the oil feed and return pipes

47 Completion of reassembly and final adjustments

1 Refit the kickstart and gearchange levers onto their respective splines, make sure they are at the correct angle and tighten the pinchbolts. Check the throttle and choke controls for correct operation noting that the carburettor slide drops fully down when the twistgrip is closed. If a decompressor valve (valve lifter) is fitted check for satisfactory operation.

2 Refill the engine oil tank to the full mark on the dipstick or, on earlier models to the level mark on the outside of the tank. Do not overfill as the correct airspace within the tank is essential to ensure correct breathing of the engine. Pour ½ pint of clean engine oil into the cran case sump, this can be done by removing the tappet inspection cover on the rocker box and using a funnel.

3 Refill the gearbox with EP 90 oil to the full mark shown on the dipstick. On earlier 250 and 350 cc Star models a stand pipe is fitted through the drain plugs to check the gearbox oil level remove the small screw in the centre of the plug and fill the box until oil drips from the screw hole; then replace the screw.

4 Remove the inspection cap at the rear of the primary chaincase and the highest of the two red painted screws at the base of the chaincase. Fill with SAE 20 W oil until it runs out of the level holes and replace inspection cap and screw.

5 Check the electrolyte level and specific gravity in the battery and if correct replace in machine and connect terminals.

6 A few of the machines supplied for both military and Government use do not have the conventional filler and drain plug in the primary chaincase. On these models the chaincase is drained by removing the lowest cover retaining screw which normally has a red painted head. The clutch adjustment orifice is used for refilling the chaincase.

48 Starting and running the rebuilt engine

1 When the initial start-up is made, run the engine slowly for the first few minutes, especially if the engine has been rebored. Check that all the controls function correctly and that there are no oil leaks.

2 With the engine idling, remove the oil tank filler cap and check that the oil from the engine sump is being pumped back into the tank. During the first few minutes of running the engine may emit a certain amount of smoke from the exhaust pipe - this is probably due to excessive oil in the sump and is not serious as the pump will quickly scavenge the oil back into the tank and the engine should stop smoking.

3 Before taking the machine on the road check the tension of the rear chain and, if fitted, ensure that the rear chain oiler at the back of the primary chaincase is slowly dripping oil onto the chain.

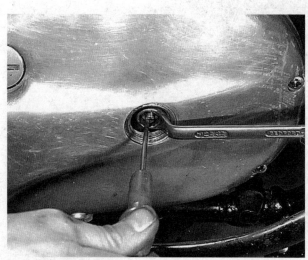

46.2 Adjusting the clutch

47.1a Adjust the gearchange lever ...

47.1b ... and footrest to the most convenient position

47.1c Replace throttle slide in the correct position

47.2 Refill the engine oil tank (approximately 4 pints)

47.3 Fill the gearbox to the correct level on the dipstick (later models)

47.4 The primary chaincase must also be refilled

49 FAULT DIAGNOSIS - Engine

Symptom	Reason/s	Remedy
Engine will not start	Defective spark plug	Remove plug and lay on cylinder head. Check whether spark occurs when engine is kicked over.
	Dirty or closed contact breaker points	Check condition of points and whether gap is correct.
	Battery terminals loose or corroded	Clean and tighten.
	Ignition coil defective	Remove HT lead from plug and jam between cylinder head fins, take off C/B cover, switch on ignition and using an insulated screwdriver flick the C/B points open — a fat consistent spark should result.
	Fuel starvation	Check fuel supply. Check to see that the fuel tap is turned on. Check the fuel filter and lines for obstruction.
	Fuel flooding	Remove and dry the spark plugs.
	Low compression	If the engine can be turned over on the kick-starter with less than normal effort, perform a compression test and determine the cause of low compression. (To check compression buy or borrow a test gauge of the type that is screwed or held in the spark plug hole while the engine is kicked over.)
Engine runs unevenly	Ignition system fault	Check system as though engine will not start.
	Blowing cylinder head gasket	Leak should be evident from oil leakage where gas escapes.
	Incorrect ignition timing	Check timing and reset if necessary.
	Loose pin on which moving contact breaker point pivots	Replace defective parts.
	Incorrect fuel mixture	Adjust carburettor air screws. Refer to Chapter 2, carburettor adjustment section. Remove and clean carburettor low speed jets. Refer to Chapter 2. Check float level. Refer to Chapter 2. Check for intake air leaks. Make sure that the carburettor mounting bolts are tight.
Lack of power	Incorrect ignition timing	See 'Ignition timing' in Chapter 1.
	Fault in fuel system	Check system and filler cap vent.
	Blowing head gasket	See above.
High oil consumption	Cylinder barrel in need of rebore and o/s piston	Fit new rings and piston after rebore.
	Oil leaks or air leaks from damaged gaskets or oil seals	Trace source of leak and replace damaged gaskets or seals.
	Oil not returning to tank	Remove sump plate and check scavenge ball valve is seating correctly. If trouble persists check oil pump.
Excessive mechanical noise	Worn cylinder barrel (piston slap)	Rebore and fit o/s piston.
	Worn small end bearing (rattle)	Replace bearing and gudgeon pin.
	Worn big-end bearing (knock)	Fit new big-end bearing.
	Worn main bearings (rumble)	Fit new journal bearings and seals.
Engine overheats and fades	Pre-ignition and/or weak mixture	Check carburettor settings. Check also whether plug grade correct.
	Lubrication failure	Check operation of pump as above, also check for possible oilway/oil pipe blockage.

Fault diagnosis - Clutch

Symptom	Reason/s	Remedy
Engine speed increases but machine does not respond	Clutch slip	Check clutch adjustment for pressure on pushrod. Also free play at handlebar lever. Check condition of clutch plate linings, also free length of clutch springs. Replace if necessary.
Difficulty in engaging gears. Gear changes jerky and machine creeps forward, even when clutch is fully withdrawn	Clutch drag	Check clutch adjustment for too much free play.
	Clutch plates worn and/or clutch drum	Check for burrs on clutch plate tongues or indentations in clutch drum slots. Dress with file.
	Clutch assembly loose on mainshaft	Check tightness of retaining nut. If loose, fit new tab washer and retighten.
Operating action stiff	Damaged, trapped or frayed control cable	Check cable and replace if necessary, Make sure cable is lubricated and has no sharp bends.
	Bent pushrod	Replace.

Fault diagnosis - Gearbox

Symptom	Reason/s	Remedy
Difficulty in engaging gears	Gear selector forks not indexed correctly	Check position of selector forks on cam plate assembly.
	Gear selector forks bent	Replace.
	Broken or misplaced selector springs	Replace broken springs and re-locate as necessary.
	Clutch drag	See 'Fault diagnosis - Clutch'.
Gears jump out of mesh	Worn dogs on ends of gear pinions	Replace worn pinions.
	Cam plate plunger stuck	Free plunger assembly.
Kickstarter does not return when engine is turned over or started	Broken or badly tensioned kickstarter return spring	Replace spring or retension.
Gear change lever does not return to normal position	Broken return spring	Replace.

Chapter 2 Fuel system

Contents

Specifications

Petrol tank capacity:

250 cc Star	2½ Imp. gallons
350 cc Star	3 Imp. gallons
250 cc B25/C25	3¼ Imp. gallons
441 cc B44	3¼ Imp. gallons
500 cc B50	3 Imp. gallons

Carburettor

250 cc Star

Type	Amal 'Monobloc'
Main jet	140
Pilot jet	N/A
Needle jet	0.1055 in.
Needle position	3
Throttle slide	4

350 cc Star

Type	Amal 'Monobloc'
Main jet	190
Pilot jet	20
Needle jet	0.1055 in.
Needle position	3
Throttle slide	376/3

250 cc B/25 - C/25

Type	Amal 'Concentric' 928/20
Main jet	200
Needle jet	0.106
Needle position	1
Throttle slide	3
Choke diameter	28 mm

441 cc B44

Type	Amal 'Concentric' 930/11
Main jet	230
Pilot jet	25
Needle jet size	0.107 in.
Needle position	2
Throttle slide	3
Nominal choke size	30 mm

500 cc B50

Type	...	...	...	...	...	...	...	...	...	Amal 'Concentric' 930/62
Main jet	...	...	...	...	...	...	...	...	...	200
Needle jet	...	...	...	...	...	...	...	...		0.106
Needle position	...	...	...	...	...	...	...			1
Throttle slide	...	...	...	...	...	...	...			3½
Choke diameter	...	...	...	...	...	...	...			30 mm

1 General description

1 The fuel system comprises a petrol tank from which petrol is fed by gravity to the float chamber of the Amal carburettor. A petrol tap with a built-in gauze filter is located beneath the rear end of the petrol tank, which has provision for turning on a small reserve quantity of fuel when the main content of the tank is exhausted. Some later models are fitted with two taps - one main and one reserve. There is an additional filter within the main feed union of both the 'Monobloc' and 'Concentric' carburettors.

2 For cold starting purposes both types of carburettor are fitted with a spring loaded air valve with the main throttle slide, the valve is controlled by the 'choke' lever on the handlebars.

2 Petrol tank removal and replacement

1 The petrol tank on all models is exceptionally easy to remove, as only one retaining nut is employed. The nut which is underneath a rubber grommet at the top, front of the tank screws down onto a stud fixed to the top tube of the frame.

2 To remove the tank disconnect the petrol feed pipe from the carburettor, undo the centre retaining nut and lift the rear of the tank up first and then withdraw it rearwards from the machine just lifting the front sufficiently to clear the retaining stud. Do not lose the rubber spacer and washers that fit on the retaining stud.

3 If the tank is badly rusted inside a method of descaling it is to obtain some ball bearings, remove the tap(s) and plug the hole(s) with tapered pieces of wood and pour the ball bearings into the tank with two or three pints of petrol. Shake the tank vigorously for as long as you can and then remove cap and empty out the bearings and petrol. When you are sure that every bearing is out flush the tank thoroughly with clean petrol several times and then replace the tap(s).

4 When replacing the tank check that the rubber mounting pads on the frame are correctly positioned and before refitting the centre retaining nut ensure that the rubber spacer and washers are replaced on the stud.

3 Petrol tap - removal and replacement

1 The petrol tap is threaded into an insert in the rear of the petrol tank. Earlier models are fitted with a single lever type tap and some models incorporate two taps with knurled buttons which are pulled out to turn the fuel on and turned anti-clockwise to lock.

2 Before the tap can be unscrewed by applying a spanner to the flats close to the petrol tank joint, the tank must first be completely drained of fuel. When the tap is removed the gauze filter will be exposed, which is an integral part of the extension of the tap body.

3 On the later models only, the petrol feed pipe is a push fit on the tube at the base of the tap. Earlier models have a screwed union joint.

4 Replace by screwing tap back into tank, checking that the fibre sealing washer is in good condition.

4 Petrol feed pipe - inspection

1 On the early models a metal feed pipe was used with screwed union joints at each end. This arrangement is unlikely to give trouble unless a blockage occurs, part of the tube is kinked or if the unions come unsoldered. Replacement will be necessary in each case, unless the blockage can be cleared or the unions re-soldered in position.

2 Later models use a plastic pipe that is a push fit on the end of the petrol tap outlet and has a union joint where connection is made with the carburettor float chamber. Replacement is necessary if the tube becomes hard or cracks, or if the push-on joint becomes slack.

5 Carburettor - removal

1 Before removing the carburettor turn off the petrol tap(s) and detach the petrol feed pipe where it joins the float chamber. Remove also the throttle slide and needle assembly by unscrewing the top of the mixing chamber and lifting it away complete with the control cable(s) and slide assembly. (Remove cross head retaining screws, 'Concentric' carburettor only).

2 On some models the air cleaner can be removed with the carburettor, on other models the clip securing the air filter hose must be unscrewed before removing the two carburettor flange nuts. Slide the carburettor off the two studs complete with the 'O' in the centre of the flange.

5.2 On B25 models, it is easier to remove the carburettor after first removing the air cleaner

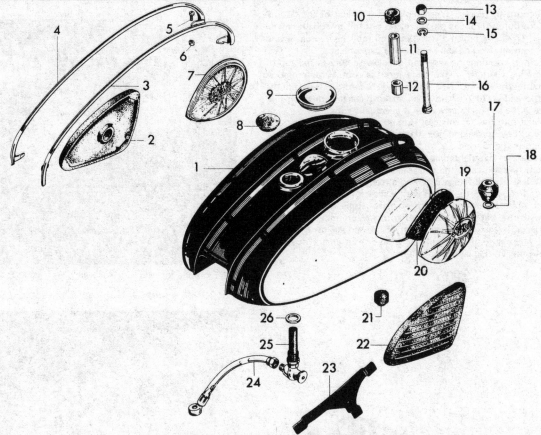

FIG. 2.1. PETROL TANK – COMPONENTS (EARLIER 'STAR' MODELS)

1 Petrol tank	8 Rubber grommet	15 Washer (spring)
2 LH knee grip	9 Filler cap	16 Centre stud
3 RH chrome trim	10 Rubber sleeve	17 Rubber grommet
4 LH chrome trim	11 Distance tube (long)	18 Washer (plain)
5 Trim retaining screw	12 Distance tube (short)	19 RH nameplate
6 Trim retaining nut	13 Tank retaining nut	20 Packing piece
7 LH nameplate	14 Washer (plain)	21 Mounting rubber

22 RH knee grip
23 Knee grip retainer
24 Petrol pipe
25 Tap filter and body
26 Fibre washer

6 Carburettor - operation

1 In simple terms, the carburettor comprises a float chamber, fuel delivery jet and a venturi tube. Fuel from the petrol tank is gravity fed into the float chamber where the level of fuel is constantly monitored by a float operated needle valve. As fuel is consumed by the engine the float drops, unseating the needle valve and allowing fuel to refill the float chamber. When the float chamber is filled to a fine-determined level the needle valve is re-seated by the rising float and fuel flow into the carburettor ceased until the level of fuel within the float chamber again falls.

2 The fuel delivery jet allows fuel from the float chamber to be metered into the engine via the venturi. The top opening of the jet is just a little bit higher than the fuel level in the float chamber. This keeps fuel always present in the jet, but never free-flowing out of it.

3 At the top opening of the fuel jet is the venturi, or carburettor throat. The venturi is a passage that is narrower in the centre than at the ends. When air rushes through it due to engine suction, a low pressure area ia created in the narrow section, and in an effort to equalize this low pressure, fuel is forced, or sucked, into the venturi passage and then into the engine in the form of an atomised spray.

4 The above description regarding the component parts is of course much simplified, for cold starting purposes there is an integral air slide or 'choke' and when the engine is idling and does not require a strong fuel/air mixture a pilot jet comes into operation. The twistgrip directly controls an air slide and jet needle, the former controlling the volume of air passing into the engine and the latter the amount of fuel.

7 Carburettor - dismantling and inspection

1 The float chamber is an integral part of the 'Monobloc' carburettor fitted to the 250 and 350 Star models, and cannot be separated. In the case of the 'Concentric' carburettor, fitted to all later models, the float chamber is held to the mixing chamber by two screws with spring washers, fitted from the underside.

2 Although the 'Monobloc' float chamber cannot be detached, access to the internals is gained by removing the two slotted head screws in the side of the float chamber and detaching the cover and gasket.

3 Check the float needle and its seating for wear and whether the needle is bent. If there is a ridge around the needle and/or its seating, replacement will be needed.

4 Check the condition of the float and whether it has become porous so that petrol will leak inwards. The earlier copper floats are more prone to this type of fault. A faulty float should be replaced, for it is not practicable to effect a permanent repair.

5 Check the throttle slide for wear and make sure none of the carburettor jets are blocked. Never use wire or any other pointed instrument to clear a blocked jet, otherwise there is risk of enlarging the jet and upsetting the carburation. Either blow the jet clear or use a blast of air from a tyre pump.

6 Make sure the float chamber body is clean and free from any sediment that may have originated from the petrol. Do not forget the tiny nylon gauze that is fitted within the petrol feed union of both the 'Monobloc' and 'Concentric' carburettors. This must be clean.

7 To reassemble the carburettor, follow the dismantling

instructions in reverse. When replacing the float of the 'Mono-bloc' carburettor on its hinge pin, do not fail to add the metal spacer. If this is omitted, the float can foul the end cover and stick, causing flooding and poor engine performance.

8 Before inserting the slide and needle assembly in the top of the mixing chamber, make quite sure the control cable nipples are correctly seated. If a cable end becomes misplaced the throttle slide will be held open making starting difficult and allowing the engine to race when it eventually runs.

9 Make sure the 'O' ring seal is in good shape and is positively located, before making the flange joint, do not use gasket cement on the 'O' ring.

10 Beware of overtightening the bolts on the carburettor flange because this will cause the flange to distort and bow, resulting in an air leak. If the flange is bowed already, remove the 'O' ring seal and rub down the flange with a sheet of emery cloth (fine grade) wrapped around a piece of glass. Apply the carburettor flange to this flat surface and rub with a rotary movement until the flange is again completely flat, as viewed against a straight edge. Make sure the carburettor is free from emery dust and replace the 'O' ring before refitting.

7.5 Examine throttle slide for wear

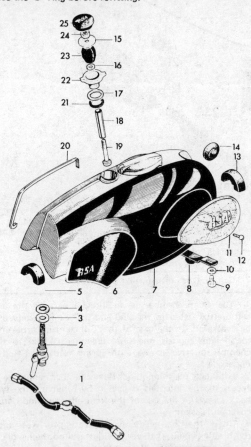

FIG. 2.2. PETROL TANK — COMPONENTS (LATER MODELS)

1 Petrol pipe	13 Mounting rubber
2 Petrol tap and filter	14 Anti-roll rubber
3 Fibre washer	15 Washer - large
4 Sealing washer	16 Washer - small
5 Mounting rubber	17 Mounting ring
6 Knee grip - 2 off	18 Distance tube
7 Petrol tank	19 Centre stud
8 Tank strap	20 Chrome trim
9 Strap bolt	21 'O' ring
10 Strap washer	22 Beading holder
11 Nameplate - 2 off	23 Mounting rubber
12 Nameplate screw - 2 off	24 Tank retaining nut
	25 Rubber grommet

7.8 Check condition of throttle slide nipple

8 Carburettor - replacing and checking the settings

1 When replacing the carburettor refrain from over-tightening the retaining bolts at the flange joint. The bolts should be fitted with spring washers, to prevent them from slackening off.

2 In the case of the 'Monobloc' carburettor, check that the small distance piece is inserted over the float hinge pin, after the float has been replaced. This small part is easily lost or mis-placed, yet if it is omitted, the float can work along the hinge pin and rub on the end plate which will limit its travel and cause the carburettor to flood.

3 It is easy to cross-thread the carburettor top on the 'Monobloc' carburettor when it is re-engaged with the car-burettor body, due to the very fine threads used. If the top is cross-threaded, an air leak may occur which will upset the carburation.

4 Always use a new 'O' ring, for the flange joint, even if the original appears to be in reasonable condition. This will eliminate one possible source of an air leak.

5 The sizes of the jets, throttle slide, needle and needle jet are

predetermined by the manufacturer and should not require modification unless the engine has been tuned. Check with the Specifications list if there is any doubt about the values fitted.

6 Slow running is controlled by a combination of throttle stop and pilot jet settings. Commence by screwing in the throttle stop screw so that the engine runs at a fast tick-over speed. Adjust the pilot jet screw until the tick-over is even, without either misfiring or hunting. Unscrew the throttle stop screw until the desired tick-over speed is obtained and check by turning the pilot jet screw either way until the tick-over is even. Always make these adjustments with the engine at normal working temperature. The normal setting for the pilot jet screw is about 1½ full turns open

from the closed position. Note that the mixture is enriched by screwing the pilot jet screw inwards on both types of carburettor.

7 As a rough guide, up to 1/8th throttle is controlled by the pilot jet, from 1/8th to 1/4 throttle by the throttle slide cutaway, from 1/4 to 3/4 throttle by the needle position and from 3/4 to full throttle by the size of the main jet. These are only approximate divisions; there is a certain amount of overlap.

8 The level of fuel in the float chamber is critical and is fine set when the carburettor is manufactured. On no account should the float hinge arm on the 'Monobloc' carburettor be bent in an effort to change the level of fuel.

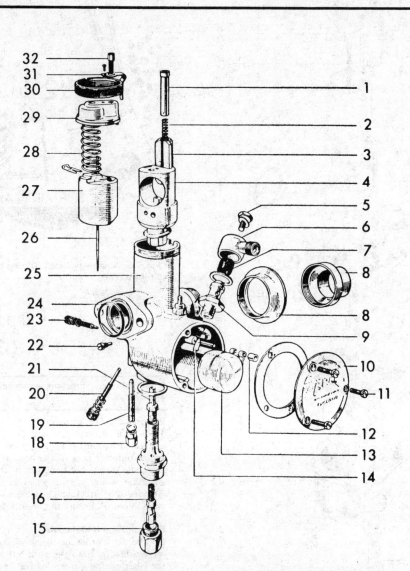

FIG. 2.3. COMPONENT PARTS OF THE 'MONOBLOC' CARBURETTOR

1 Air valve guide
2 Air valve spring
3 Air valve
4 Jet block
5 Banjo bolt
6 Banjo
7 Filter gauze
8 Air filter connection (top) or air intake tube
9 Needle setting
10 Float chamber cover
11 Cover screw
12 Float spindle bush
13 Float
14 Float needle
15 Main jet cover
16 Main jet
17 Main jet holder
18 Pilot jet cover nut
19 Pilot jet
20 Throttle stop screw
21 Needle jet
22 Locating peg
23 Air screw
24 'O' ring seal
25 Mixing chamber
26 Jet needle
27 Throttle slide
28 Throttle spring
29 Top
30 Cap
31 Click spring
32 Adjuster

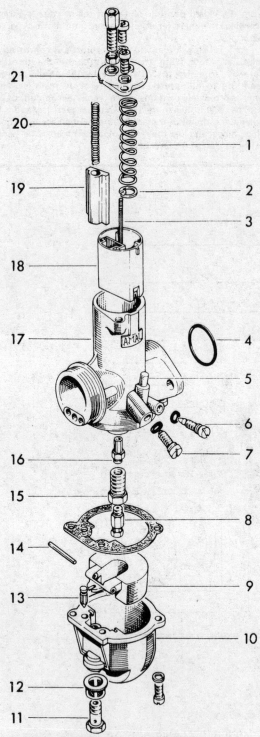

FIG. 2.4. COMPONENT PARTS OF THE
CONCENTRIC CARBURETTOR

1	Throttle spring	11	Banjo bolt
2	Needle clip	12	Filter
3	Throttle needle	13	Float needle
4	'O' ring	14	Float spindle
5	Tickler	15	Jet holder
6	Pilot air screw	16	Needle jet
7	Throttle stop	17	Carburettor body
8	Main jet	18	Throttle valve
9	Float	19	Air slide
10	Float chamber body	20	Air slide spring
		21	Mixing chamber cap

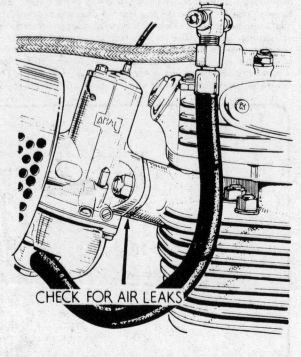

CHECK FOR AIR LEAKS

Fig. 2.5. A wrongly fitted 'O' ring or distorted flange can cause air leaks here

9 Air cleaner - location, examination and replacement of the element

250 and 350 Star models:

1 On the earlier 250 and 350 machines the air cleaner is concealed within the fairing below the seat and is connected to the carburettor by a rubber hose. Remove the two nuts retaining the cover on the left-hand side of the machine provides access to the filter which is retained in its housing by a spring-wire clip. Remove the filter, which is the wire gauze and fabric type and wash in petrol. Ensure the filter is dry before replacing it.

B25, C25 and B44 models:

2 The air cleaner on these models is the pill-box type and screws directly onto the carburettor air intake. After unscrewing the filter can be dismantled for cleaning, by unscrewing the bolt holding the perforated outer cover together.

3 The element is made of surgical gauze and should be washed thoroughly in petrol and allowed to dry before replacement. If the element appears unserviceable, it should be renewed.

B50 models:

4 The 500 cc models are fitted with a paper element type filter located behind the panel on the left-hand side of the machine. The element is retained by a single centre nut and washer and after removal, can be cleaned with a stiff brush or, if available, an air line. Do NOT wash in petrol or detergent. To maintain the engine at peak performance the element should be renewed every 5000 miles.

9.2 'Pill box' type air cleaner unscrews from carburettor intake

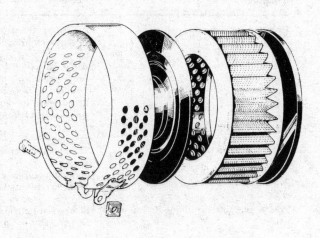

Fig. 2.7. Dismantling 'pill box' air cleaner

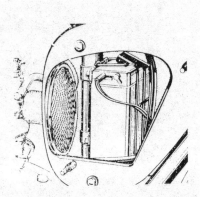

Fig. 2.6. Location of intake filter on 250 and 350 'Star' models

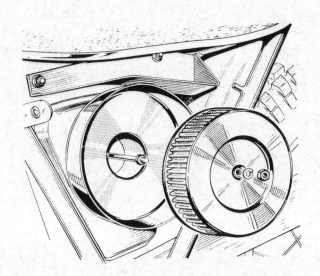

Fig. 2.8. Later type replaceable paper filter

10 Exhaust system

Unlike a two-stroke, the exhaust system is unlikely to require attention during the normal service life of the machine, other than periodic replacements. Always replace the silencer with one of the original design or with one expressly designed for the BSA twins. Although a change of silencer may give the impression of greater speed as the result of the changed exhaust note (often louder) in a great many cases it has been found that performance suffers as a result. Many mysterious power losses and rough running can be attributed to an unwitting change of silencer, when even juggling with the jet sizes of the carburettor will give only an uneasy compromise.

11 FAULT DIAGNOSIS – FUEL SYSTEM AND CARBURATION

Symptom	Reason/s	Remedy
Excessive fuel consumption	Air cleaner choked or restricted	Clean or if paper element fitted, replace.
	Fuel leaking from carburettor. Float sticking	Check all unions and gaskets. Float needle seat needs cleaning.
	Badly worn or distorted carburettor	Replace.
	Jet needle setting too high	Adjust as figure given in specifications.
	Main jet too large or loose	Fit correct jet or tighten if necessary.
	Carburettor flooding	Check float valve and replace if worn.
	Air slide (choke) remaining partially closed	Check slide return spring and lubricate/renew as necessary.
Idling speed too high	Throttle stop screw in too far. Carburettor top loose	Adjust screw. Tighten top.
	Pilot jet incorrectly adjusted	Refer to relevant paragraph in this Chapter.
	Throttle cable sticking	Disconnect and lubricate or replace.
Engine dies after running for a short while	Blocked air hole in filler cap	Clean.
	Dirt or water in carburettor	Remove and clean out.
General lack of performance	Weak mixture; float needle stuck in seat	Remove float chamber or float and clean.
	Air leak at carburettor joint	Check joint to eliminate leakage, and fit new 'O' ring.
Engine does not respond to throttle	Throttle cable sticking	See above.
	Petrol octane rating too low	Use higher grade (star rating) petrol.

Chapter 3 Ignition system

Contents

Specifications

250 and 350 Star models

Battery	6 volts
Alternator	
Make	Lucas
Output	Rectified DC for lighting and battery charging, also provides an emergency start facility flat (battery)

B25, C25, B44 and B50 models

Battery	12 volts
Alternator	
Make	Lucas
Output	Rectified DC for lighting and battery charging via a Zener diode. High alternator output provides flat battery starting
Contact breaker gap	0.015 inch - all models
Spark plug	
250 and 350 Star series	Champion N5
B25 and C25 series	Champion N3
B44 and B50 series	Champion N4
Plug gap	0.020 — 0.025 inch - all models

1 General description

1 All the BSA single cylinder models covered in this manual are equipped with battery and coil ignition, the battery being charged from an alternator, (AC generator) driven from the primary drive side of the crankshaft. The high tension (HT) current from the coil to the spark plug is initiated and controlled by a low tension (LT) contact breaker unit driven from the camshaft and located on the timing cover. On earlier Star models the CB unit is driven from the oil pump pinion and is located vertically behind the cylinder barrel.

2 The LT circuit comprises basically of the battery lead to the ignition switch, a lead from the ignition switch to the primary coil windings (terminal SW on the coil) and finally, a lead from the coil secondary windings, (terminal CB) to the condenser and points within the contact breaker assembly.

3 High tension (HT) voltage is generated in the coil when the low tension circuit is broken by the opening of the contact breaker points, the HT current passes along the thick HT lead to the centre electrode of the spark plug and jumps to earth on the body of the plug inside the combustion chamber, creating a high intensity spark. When the engine is running or being kicked over, this spark is sufficient to ignite the highly explosive mixture of compressed petrol vapour and air inside the combustion chamber.

4 On earlier models, the ignition switch or knob has an emergency start position that can be used if the battery is flat. When the emergency position is selected, the alternator output passes directly to the ignition coil and provides sufficient current for starting, the battery also receives a charging current. As soon as the engine has started the ignition switch must be returned to the normal running position to provide the battery with the correct charging current. Failure to do this may cause the engine to misfire and possibly damage the battery.

5 Later BSA machines are equipped with 12 volt lighting and ignition systems and the alternator output is sufficiently high to enable the engine to be started with a completely flat battery without the use of an emergency start circuit in the ignition switch.

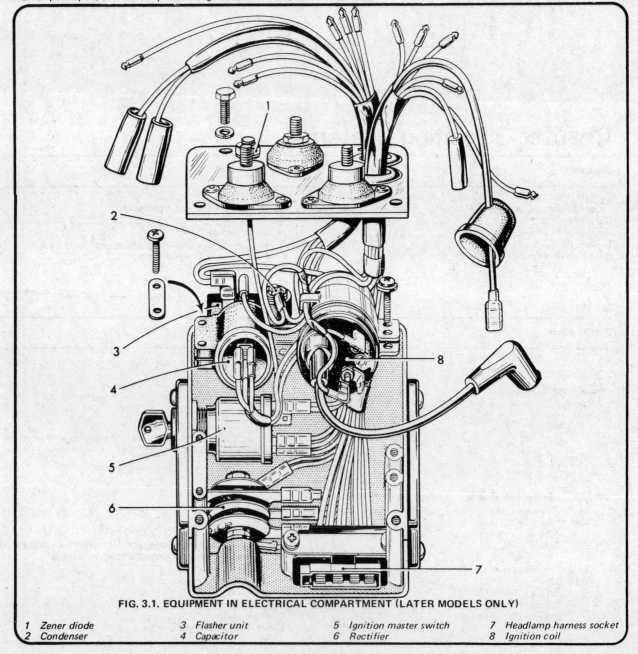

FIG. 3.1. EQUIPMENT IN ELECTRICAL COMPARTMENT (LATER MODELS ONLY)

| 1 | Zener diode | 3 | Flasher unit | 5 | Ignition master switch | 7 | Headlamp harness socket |
| 2 | Condenser | 4 | Capacitor | 6 | Rectifier | 8 | Ignition coil |

2 Capacitor ignition system - 1971 and later models only

1 1971 and later BSA singles are fitted with a capacitor in addition to the battery, to enable the machine to be used for competition purposes with the battery removed.

2 The capacitor is located alongside the coil in the electrical compartment beneath the fuel tank. If it is intended to use the capacitor and the battery is removed: prevent the negative battery cable (brown and blue) from earthing on the frame by wrapping some insulation tape around it.

3 To maintain the capacitor in good condition, the machine should be run for a few minutes with the battery disconnected twice a year. Do not run the engine with the zener diode disconnected, or the capacitor will be damaged due to the excessive voltage. If you experience hard starting with the battery connected, disconnect the capacitor to eliminate the possibility of a short circuit.

3 Capacitor - checking

1 The capacitor has a limited storage life of approximately 18 months at 68°F when not being used, therefore it is a good policy to remove the battery and run the engine using the capacitor only, at least twice a year to check that it is operational. If the engine will not start or misfires when the capacitor is in circuit it can be checked as follow:-

2 Make a careful note to which cable goes on which capacitor terminal; the single terminal is marked with a red dot and is the positive (earth) connection. The double terminal is the negative connection.

3 After checking as detailed above, remove the cables, slacken the two clamping screws and slide out the capacitor.

4 Connect the machines 12v battery directly to the capacitor (negative to blue/brown terminal, positive to red terminal) for five seconds.

5 Disconnect the battery and let the capacitor stand for five minutes.

6 Buy, or preferably borrow a DC voltmeter and connect it across the capacitor terminals. If the capacitor is functioning properly, the voltmeter will read not less than 9 volts after the needle has steadied.

4 Ignition coil - checking

1 The ignition coil is a sealed unit, designed to give long service. On earlier models it is located beneath the top frame tube and to gain easier access to it the fuel tank must be removed first (see Chapter 2). On later models the coil is housed in the electrical compartment beneath the fuel tank.

2 To test the coil, first ensure that the terminals are clean and tight and the HT lead is correctly connected within the coil. Check that the battery is fully charged and remove the contact breaker dust cover, turn the engine over slowly until the points are closed.

3 Wedge the spark plug end of the HT lead between two cylinder barrel fins so that the bared end of the centre wire is approximately 3/16'' - ¼'' away from the metal of the cooling fins.

   Switch on the ignition and, using a plastic handled screwdriver flip the contact breaker points open - a healthy and quite audible spark should jump from the end of the HT lead to the cylinder barrel fins. Repeat the operation several times, then switch off the ignition to avoid damage to the coil.

4 If no spark results and it is known that the battery, condenser and contact breaker points are not at fault, (see following paragraphs) take it to an electrical repair expert for checking. A faulty coil must be replaced as it is not practicable to effect a repair.

4.1 Location of ignition coil on earlier models

5 Contact breaker - adjustment

1 On the earlier 'Star' machines manufactured before 1965 the contact breaker unit is mounted above the crankcase behind the cylinder barrel. The cover is held in place by a spring clip or, on later models a central screw. The innermost contact point is fixed to a sliding plate, to adjust the gap, remove the sparking plug and rotate the engine until the points are in the fully open position. Slacken the sliding plate retaining screw and move the plate either backwards or forwards to obtain the correct gap at the points.

2 In 1965 the vertical type CB unit was discontinued, being replaced by a unit within the timing cover driven from the end of the camshaft. Access is gained by removing the two screws retaining the cover onto the timing chest. To adjust the points, first set them to the maximum open position as explained above and then slacken the single slotted locking screw shown in the adjacent diagram and set the points to the required gap. Tighten the locking screw and recheck the gap.

3 Models manufactured after 1967 have a similar unit to that used on preceding models with the exception that the adjustment procedure is extremely simple and more accurate. First the locking screw, (see accompanying diagram) is slackened and the eccentric pin below it is turned either way to obtain the correct gap. Retighten the locking screw and check the gap has not altered. It is important on all models that the contact points are in the fully open position before attempting to adjust the points, otherwise a completely false setting will result. If the gap is correct the feeler gauge should be a good sliding fit.

4 Before replacing the cover, place a very slight smear of grease on the contact breaker cam or one or two drips of light oil on the felt pad that presses on the cam (if fitted).

6 Contact breaker points – removal, renovation and replacement

1 If the contact breaker points are burned, pitted or badly worn, they should be removed for dressing. If it is necessary to remove a substantial amount of material before the faces can be restored, the points should be replaced.

2 To remove the points on earlier models fitted with the vertical CB unit slacken the low tension wire terminal nut on the outside of the housing and lift the spring holding the outer point from the clamp, remove the adjusting nut from the sliding plate and withdraw the lower contact point.

3 On models manufactured between 1965 and 1967 it is necessary to remove the centre locking screw and the condenser retaining screw and slide both points and condenser off together. The condenser can then be detached easily from the contact breaker spring.

4 To remove the points on all models manufactured after 1967, first remove the locking screw, unfasten the low tension wire terminal nut and remove the insulating washers underneath making careful note of their positions on the spigot. Slide the moving contact point and spring off first and then remove the small plate holding the fixed point.

5 The points should be dressed with an oilstone or fine emery cloth. Keep them absolutely square during the dressing operation, otherwise they will make angular contact when they are replaced and will quickly burn away.

6 Replace the contacts by reversing the dismantling procedure. Take particular care when replacing the insulating washers, to make sure they are fitted in the correct order. If this precaution is not observed the points will be isolated electrically and the ignition system will not function.

7 Condenser - removal and replacement

1 A condenser is included in the contact breaker circuitry to prevent arching across the contact breaker points as they separate. It is connected in parallel with the points and if a fault develops the ignition system will not function correctly.

2 If the engine is difficult to start or if misfiring occurs, it is possible that the condenser has failed. To check, separate the contact breaker points by hand whilst the ignition is switched on. If a spark occurs across the points and they have a blackened or burnt appearance, the condenser can be regarded as unserviceable.

3 It is not possible to check the condenser without the necessary test equipment. In view of the low cost of a replacement it is preferable to change the condenser and observe the effect on engine performance.

4 To remove the condenser, unscrew the terminal nut on the end and lift off the contact breaker return spring, (on earlier 'Star' models remove the flat metal strip). Remove the retaining screw and lift out the condenser. On more recent models the condenser is housed in the electrical box beneath the fuel tank and is removed by disconnecting the terminal nut and unfastening the retaining screw.

6.2 Dust cover supports lock contact breaker base plate in position

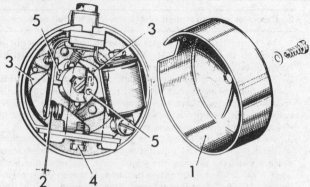

FIG. 3.2. CONTACT BREAKER ASSEMBLY AND AUTO-ADVANCE UNIT ON THE O.H.V. 250 STAR MODEL PRIOR TO THE 1965 SEASON
1   Contact breaker unit cap
2   Contact breaker points gap (0.015 inch)
3   Advance unit bob-weight
4   Adjusting screw
5   Bob-weight pivots (oil lightly)

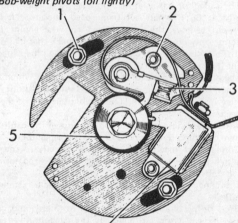

FIG. 3.3. CONTACT BREAKER ASSEMBLY ON MODELS FROM 1965 TO 1967
1   Plate securing nut   4   Condenser
2   Adjusting screw   5   Cam
3   Contact breaker points

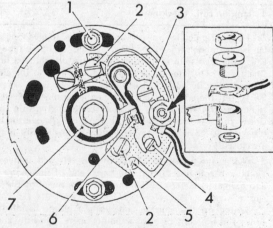

FIG. 3.4. CONTACT BREAKER UNIT – 1968 MODELS ONWARDS
1   Plate securing nut
2   Advance/retard fine adjustment retaining screws
3   Adjustment locking screw
4   Eccentric contact adjuster screw
5   Advance/retard eccentric adjuster screw
6   Contact breaker points
7   Cam

8 Ignition timing

1 Full details of ignition timing are given in Chapter 1, but the following Table 1 and associated text will provide an additional and perhaps more accurate method of timing.

Table 1 - Ignition settings

Model	Crankshaft degrees before TDC	
All 'Star' models	33½ deg.	Fully advanced
B25/C25 models	37 deg.	Fully advanced
B44 models	28 deg.	Fully advanced
B50 models	34 deg.	Fully advanced

2 Obtain a degree disc from a BSA dealer and attach it to the alternator end of the crankshaft after removing the primary drive chain cover (left-hand side) and the nut that retains the alternator rotor.

3 A pointer should then be attached to one of the tapped holes used for the outer cover retaining screws and held firmly in position, so that it points to zero when the piston is exactly at top dead centre. Adjust by rotating the timing disc on the crankshaft and then lock the disc in position with the crankshaft rotor nut.

4 Disconnect and remove the battery from the machine. Attach one battery lead to the moving contact return spring and the other, via a 6 or 12 volt bulb (depending on battery voltage) to any convenient earthing point on the machine. As soon as the contacts separate, the bulb will be extinguished.

5 Set the points so that the points are just separating and the bulb goes out. If the ignition timing is correct, the pointer should show the correct figure in degrees given in Table 1 on the disc with the contact breaker cam held in the fully advanced position.

6 On the later models, an extremely accurate method of checking the ignition timing can be made by the use of a Strobe lamp. It is unlikely that the average owner possesses one of these lamps, but it should be possible to borrow one from an engineer friend or local garage.

7 Remove the cover at the front of the primary chaincase and, using a small brush apply a dab of white paint on the rotor timing mark and the pointer on the edge of the chaincase. Connect the strobe light according to the manufacturers instructions; usually one lead is attached to the spark plug and other to the HT lead. Start the engine and aim the light toward the alternator rotor and increase the engine speed to above 4300 rpm (B25). or 3000 rpm (B44 and B50). At this point the ignition should be fully advanced and the timing pointer and mark should be aligned. If adjustment is necessary, loosen the screws and rotate the points baseplate in the required direction until the pointer and mark are aligned.

If the advance unit is unsteady or if full advance comes in at a significantly lower or higher speed that that specified, remove the breaker plate assembly and check the advance units for broken or weak springs and for stiff pivots.

8 On early models that do not have a timing pointer and mark, the engine can be set at the correct position for timing adjustment (full advance) by locating the flywheel indentation with the special threaded locating tool shown in the adjacent diagram, (BSA tool No 60 - 1859) after the crankcase plug at the front of the engine has been removed.

9 Automatic advance unit

1 All BSA singles are fitted with a centrifugal ignition advance unit forming an integral part of the contact breaker cam drive. The unit comprises a baseplate on which are mounted two pivoted weights and their associated retaining springs. When the engine is stationary, the governor springs hold the bob-weights in the inner position, where the timing is fully retarded. As the engine speed increases, the bob-weights move outwards under

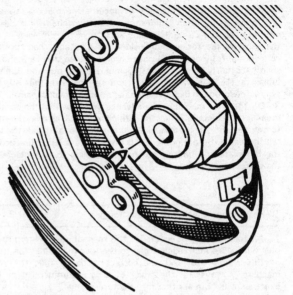

Fig. 3.5. Engine timing aperture showing the rotor timing mark and pointer (later models only)

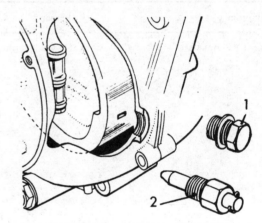

FIG. 3.6. METHOD OF IGNITION TIMING ON EARLIER MODELS

1 Blanking plug
2 Special BSA tool No. 60-1859

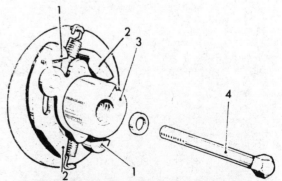

FIG. 3.7. AUTOMATIC ADVANCE MECHANISM

1 Bob-weight pivot   3 Cam
2 Bob-weight   4 Securing bolt

the influence of centrifugal force, so progressively advancing the ignition timing, until it is fully advanced at normal road speeds.

2 On the earlier Star engines the advance mechanism is clearly visible when the contact breaker dust cover is removed, check the cam spindle for excessive sideways play and then turn the cam in the fully advanced position and release, the cam should snap back fairly smartly to its original retarded position. Lightly lubricate the bob-weight pivots and springs with thin oil, use only a few drops to avoid fouling the contact breaker points.

3 On the later machines the advance unit is behind the contact breaker baseplate within the timing cover. Remove the two screws and lift out the baseplate taking care not to break the low tension cable. Before removal it is essential to scribe a line across the baseplate and housing to enable the plate to be replaced in the correct position.

10 Spark plug - checking and resetting gap

1 All the BSA singles covered in this manual use a 14 mm spark plug with a 3/4 inch reach. Refer to the specification at the front of this chapter to find the correct grade of plug for your machine. Always use the grade of plug recommended or the exact equivalent in another manufacturers range.

2 Check the gap at the plug points every 2000 miles. To reset the gap, bend the outer electrode closer to the central electrode and check that a 0.023 in feeler gauge can be inserted. Never bend the central electrode, otherwise the insulator will crack, causing engine damage if particles fall in whilst the engine is running.

3 The condition of the spark plug electrodes and insulator can be used as a reliable guide to engine operating conditions, with some experience. See accompanying illustrations.

4 Always carry at least one spare spark plug of the correct grade. This will serve as a get-you-home means if one of the sparking plugs in the engine should fail.

5 Never overtighten a spark plug, otherwise there is risk of stripping the threads from the cylinder head, especially in the case of one cast in light alloy. A stripped thread can be repaired by using what is known as a 'Helicoil' thread insert, a low cost service of cylinder head reclaimation that is operated by many dealers.

6 Use a sparking plug spanner that is a good fit, otherwise the spanner may slip and break the insulator. The plug should be tightened sufficiently to seat firmly on its sealing washer.

7 Make sure the plug insulating caps are a good fit and free from cracks. The caps contain the supressors that eliminate radio and TV interference; in rare cases the suppressors have developed a very high resistance as they have aged, cutting down the spark intensity and giving rise to ignition problems.

11 FAULT DIAGNOSIS - Ignition system

Symptom	Reason/s	Remedy
Engine will not start	No spark at plug	Try replacement plug if gap correct.
		Check whether contact breaker points are opening and closing, also whether they are clean.
		Check whether points arc when separated. If so, replace condenser.
		Check ignition switch and coil.
		Battery fully discharged. Switch off all lights and use emergency start facility.
Engine starts but runs erratically	Intermittent or weak spark	Try replacement plug. Check whether points are arcing. If so, replace condenser.
		Check accuracy of ignition timing.
		Low output from alternator causing flat battery, or imminent breakdown of ignition coil.
		Plug has fouled. Fit replacement.
		Plug lead insulation breaking down. Check for breaks in outer covering, particularly near frame. Check security of HT lead on coil and the two LT terminals.

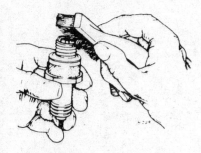

Cleaning deposits from electrodes and surrounding area using a fine wire brush.

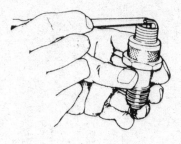

Checking plug gap with feeler gauges

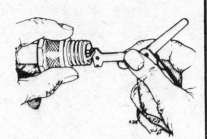

Altering the plug gap. Note use of correct tool.

Spark plug maintenance

White deposits and damaged porcelain insulation indicating overheating

Broken porcelain insulation due to bent central electrode

Electrodes burnt away due to wrong heat value or chronic pre-ignition (pinking)

Excessive black deposits caused by over-rich mixture or wrong heat value

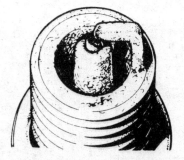

Mild white deposits and electrode burnt indicating too weak a fuel mixture

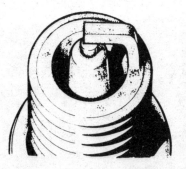

Plug in sound condition with light greyish brown deposits

Fig. 3.8. Example of spark plug conditions

Chapter 4 Frame and forks

Contents

Specifications

Frame

Type - All models	Single downtube frame with hydraulically damped front forks and swinging arm rear suspension

Front forks

Free spring length	B50 Series - main spring, 19.5 in. recoil spring .94 in. B44 and B25 Series - main spring, 10.75 in.
Oil capacity	All models - 1/3 pint
Oil viscosity	All models - SAE 20

Rear suspension

Suspension units	'Star' models - Girling non-adjustable with spring shroud B25, B44 and B50 Series - Girling adjustable with exposed chromium plated springs
Swinging arm pivot bearings	'Star' models - plain phospher bronze bushes Later 350 'Star' models and all models from 1966 until 1970 - 'Silentbloc' rubber bushes 1971 and later models - needle roller bearings

1 General description

All the BSA unit construction models are fitted with hydraulically damped telescopic front forks and swinging arm rear suspension. The cycle frame is of single downtube construction, although obviously the larger capacity machines employ heavier gauge tubing than the smaller bikes. Three types of front forks are used; the earlier models had rod type internal damping and the later design incorporated a shuttle valve for the same function. The latest forks are fitted with internal coil springs and are much slimmer and more streamlined in appearance. The 'Star' models utilized phosper bronze bushes in the swinging arm pivot bearings, later superseded by rubber bonded bushes as used on the Shooting Star, Starfire and Barracuda models. All machines manufactured during 1971 and after were further improved by the introduction of needle rollers for the swinging arm bearings. The rear suspension shock absorbers are virtually identical on all models with the exception that the 'Star' series were equipped with a metal shroud over the spring. Some earlier models were fitted with a steering damper, which, providing it is used sensibly, is a worthwhile addition.

72

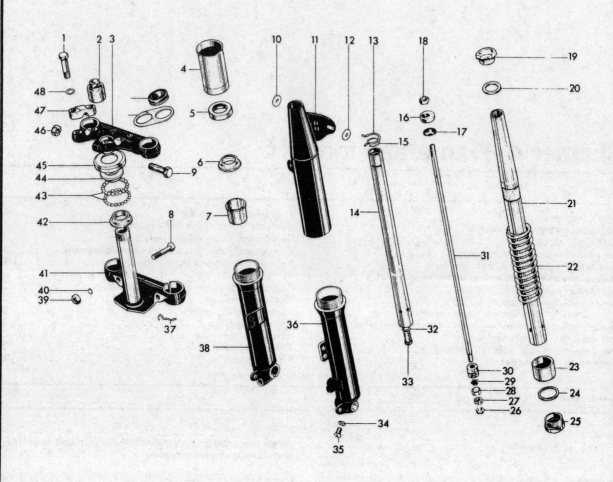

FIG. 4.1. EXPLODED VIEW OF THE ROD DAMPER TYPE FRONT FORKS AND STEERING HEAD

1 Bolt	13 Circlip	25 Plug	38 Bottom tube (LH)
2 Cap	14 Damper tube	26 Nut	39 Nut
3 Top yoke	15 Circlip	27 Seating	40 Lockwasher
4 Oil seal holder	16 Sealing washer	28 Damper valve	41 Bottom yoke
5 Oil seal	17 Retainer	29 Collar	42 Cone (lower)
6 Spacer	18 Nut	30 Bush	43 Ball
7 Bush (upper)	19 Nut	31 Damper rod	44 Cone (upper)
8 Bolt	20 Washer	32 Washer	45 Cap
9 Bolt	21 Fork shaft	33 Bolt	46 Nut
10 Spacer	22 Spring	34 Washer	47 Clamp
11 Sleeve (right hand)	23 Bush (lower)	35 Plug	48 Washer
11 Sleeve (left hand)	24 Washer	36 Bottom tube (RH)	49 Grommet
12 Washer		37 Clip	50 Cable guide bracket

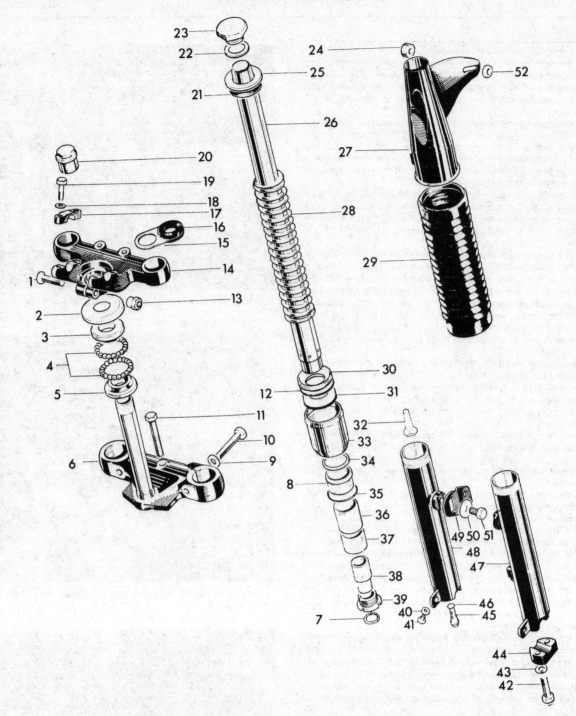

FIG. 4.2. EXPLODED VIEW OF THE SHUTTLE TYPE FRONT FORKS FITTED TO MACHINES PRIOR TO 1970

1 Pinch bolt, top yoke	14 Fork top yoke	27 Fork sleeve (RH)	40 Drain plug washer
2 Dust cap, top cone	15 Cable guide bracket	28 Fork spring	41 Fork drain plug
3 Top cone, steering head	16 Grommet, cable guide bracket	29 Fork bellows	42 Bolt, fork end cap
4 Steel ball, steering head	17 Handlebar clamp	30 Spring support	43 Spring washer, end cap bolt
5 Bottom cone, steering head	18 Washer, clamp bolt	31 Fork oil seal	44 Fork end cap
6 Fork bottom yoke	19 Bolt, handlebar clamp	32 Fork restrictor	45 Flanged bolt, restrictor
7 Shuttle valve circlip	20 Steering stem cap	33 Fork oil seal holder	46 Washer, restrictor bolt
8 Fork top bush	21 Sealing washer	34 Outer washer	47 Sliding tube (RH)
9 Washer, bottom yoke bolt	22 Washer, fork top nut	35 Fork 'O' ring	48 Sliding tube (LH)
10 Bolt, bottom yoke	23 Top nut, fork shaft	36 Fork inner sleeve	49 Mudguard bracket (RH)
11 Bolt, heat sink	24 Headlamp spacer	37 Fork lower bush	50 Washer, bracket bolt
12 Support washer, fork spring	25 Retainer, sealing washer	38 Shuttle valve	51 Bolt, mudguard bracket
13 Nut, top yoke bolt	26 Fork shaft	39 Bottom nut, fork shaft	52 Washer, headlamp mounting

2 Front forks - removal from frame

1 It is unlikely that the front forks will need to be removed from the frame as a complete unit, unless the steering head bearings require attention or the forks are damaged in an accident.

2 Commence operations by placing a stout wooden box below the crankcase to raise the front wheel off the ground approximately 6 inches. Disconnect the front brake cable by removing the split pin from the clevis pin on the cable retaining fork and detaching the outer cable. On some models it is necessary also to detach the front brake torque arm.

3 On machines fitted with a threaded wheel spindle, remove the clamping bolt at the bottom of the left hand fork and, using a tommy bar unscrew the spindle from the right hand fork and withdraw it from the hub while supporting the weight of the wheel.

4 If the spindle is retained by clamps remove the bolts (four on the larger machines), from the extreme end of both fork legs so that the bottom half of the split clamp arrangement can be detached. The front wheel will now be released from the fork ends, complete with brake plate and spindle. It may be necessary to turn the forks at an angle so that the front wheel can be lifted clear away from the mudguard. Some models are fitted with a nut on each end of the spindle, these must be slackened before the clamps are removed.

5 Detach the controls from the handlebar by either disconnecting the cable ends or by detaching the controls with the cables still attached. This includes the cut-out button, horn button, dip switch and, on later models the indicator lights switch. In the case of the horn button, it is preferable to detach the main lead from the battery before the button is removed, to prevent short circuits. Detach the speedometer drive cable by unscrewing the gland nut from the bottom of the speedometer head and the pilot bulb within the speedometer head. If a tachometer is fitted, remove in the same manner as the speedometer.

6 Remove the handlebars by withdrawing the bolts that retain the split mounting clamps to the fork top yoke.

7 Remove the headlamp by unscrewing the two retaining bolts in each side of the shell. If a cowl is fitted, this must be removed with the headlamp. The headlamp can be left to hang in a position where it is not liable to suffer damage. On some earlier 'Star' models, the headlamp nacelle is integral with the fork shrouds, in which case the front of the headlamp should be removed, the speedometer drive cable disconnected and the wires detached from the headlamp switch. It is advisable to make a note of the connections, even if the colour coding of the wires aids correct replacement.

8 Drain the oil from each fork leg and unscrew and remove the two chromium plated caps at the top of each fork leg. If the forks are the rod damper type unscrew the caps just high enough to slacken the damper rod locknuts underneath before completely removing. Slacken the pinch bolt in the fork top yoke, immediately to the rear of the steering head and unscrew the chromium plated cap at the top of the steering head. Loosen the pinch bolts on the bottom yoke. If the machine is fitted with a steering damper, it will be necessary first to remove the split pin from the bottom end of the stem and unscrew the knob until it can be detached, complete with rod.

9 When the steering head cap has been unscrewed, the fork top yoke can be removed by striking it from the underside with a rawhide mallet, first one side and then the other, this should free the forks from their tapers within the top yoke. If the forks will not unstick from the top yoke tapers; obtain BSA tool No. C1-3824 and screw it into the top of the fork stanchion. Hit the tool sharply with a mallet until the stanchion is freed. The forks, complete with steering head stem can now be drawn downwards until they are completely separate from the machine. Note that the uncaged ball bearings of the steering head assembly will drop free as the cups and cones separate, necessitating some arrangement for catching them.

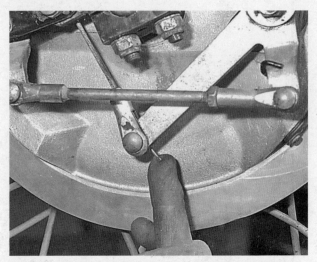
2.2 Remove split pin and clevis pin to release front brake cable

2.4 Release spindle retaining blocks to remove front wheel (later models)

2.8a Unscrew the drain plugs and ...

2.8b ... remove the stanchion caps

2.8c Slacken the bottom yoke pinch bolts

2.9 Withdrawing the fork stanchion from the top housing without removing the steering head

10 If further dismantling is necessary, the front mudguard can be removed after the forks have been separated from the frame by disconnecting the mudguard stays from the fork legs.

3 Front forks - dismantling

Shuttle Valve Type (early models)

1 First check that both legs have been drained of oil. If rubber boots are fitted slide them off the top of the forks and remove the coil springs.

2 Wrap a piece of rubber or thick cloth around the fork leg and clamp the leg gently in a soft-jawed vice.

3 Unscrew the chromium oil seal holder, using BSA service tool No. 61-6017 or a strap spanner if it cannot be turned by hand.

4 Firmly grasp the stanchion tube and move it back and forth against the top bush until the bush is driven out of the bottom section of the leg; the stanchion, complete with bushes and shuttle valve can then be removed.

5 To free the shuttle valve, remove the bottom retaining circlip and let the valve slide out the top end of the stanchion. Alternatively, unscrew the retaining plug using a 'C' spanner.

6 Do not disturb the bottom bush unless it is to be replaced. If it is worn, remove the bottom bearing retaining nut and drive the bush out with a hammer and drift. Take care not to slip and damage the stanchion tube.

7 If it is necessary to remove the restricter at the bottom of the leg, unscrew the bolt in the wheel spindle cutaway.

8 To remove the oil seals from their holders, take out the loose backing washer from the threaded end of the holder, and drive the seal out through the exposed slot. Note the 'O' ring in the threaded end of the seal holder.

Shuttle Valve Type (1971 models and later)

9 The 1971 and later machines are fitted with internally sprung forks with shuttle valve damping and are simple in design but extremely robust.

10 The internal spring and damping components can be removed from these forks without the necessity of removing the fork stanchions from the top and bottom retaining yokes. However, if the stanchions are damaged they should be removed as described in the 'Removal' Section of this Chapter.

11 First remove the fork drain screws and while the oil is draining into a tin remove the two springs from inside the fork stanchions, (it is assumed that the two top stanchion caps have already been removed).

12 Engage BSA service tool No. 61-6113, or a piece of flat steel filed to size, with the slots in the top of the damper tube and unscrew the Allen bolt at the bottom of the fork while preventing the damper tube from turning.

13 Pull the lower section of the fork leg off the main stanchion.

14 Remove the damper assembly from the bottom of the stanchion, taking care not to damage the aluminium end plug when unscrewing it.

15 Unscrew the valve retaining nut and remove the valve and shuttle washer.

16 Place the lower fork section in a vice and protect the jaws with thick cloth or pieces of wood. Collapse the seal inwards using a pointed aluminium drift, with extreme care to avoid damaging the fork leg. Prise the seal out from its housing. The small dust excluding boot will of course have to be removed prior to this operation.

Rod Damper Type

17 Drain the oil from each leg and remove boot (if fitted) and springs after first slackening lower yoke pinch bolts as detailed in the previous paragraph.

18 Clamp the fork leg in a soft-jawed vice at the spindle lug, at the bottom of the leg and slide BSA service tool No. 61-3005 over the main tube and engage the dogs at the bottom of the oil seal holder.

   While applying pressure to the end of the tool, turn it counterclockwise and free the seal holder.

19 Remove the special tool and slide the seal holder to the end of the tube, (do not attempt to entirely remove the seal holder because damage may result), the main tube assembly and lower sliding leg can now be separated.

20 Clamp the unmachined portion of the tube in a soft-jawed vice protected with thick cloth and remove the large nut at the base of the shaft. Remove the bush, spacer, and oil seal assembly.

21 Remove the socket screw that secures the damper tube to the lower portion of the fork leg.

22 Remove the two circlips at the top of the damper tube. This will free the damper rod with valve and bush.

23 Remove the nut that secures the damper valve to the rod. Do not disturb the sealing washer and special retainer located just below the nut unless they require replacement.

24 If an oil seal requires replacement, position the holder with the bottom edge on a wooden block and drive out the seal with BSA service tool No. 61-3007 or a suitable substitute.

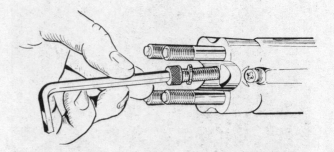

Fig. 4.5. Removing the damper valve retaining screw

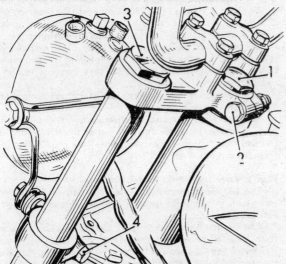

FIG. 4.3. FORKS AND STEERING HEAD RETAINING POINTS ON 1971 AND LATER MODELS (EARLIER MODELS ARE SIMILAR)

1   Steering head adjuster nut   3   Fork top cap
2   Steering head pinch bolt   4   Bottom yoke pinch bolt

3.1 Slide off the rubber gaiter to expose the coil spring (earlier shuttle type forks)

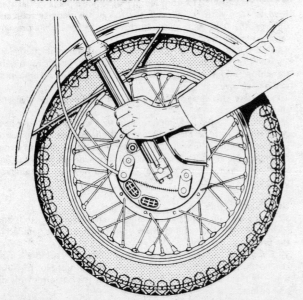

Fig. 4.4. Testing the steering head for play

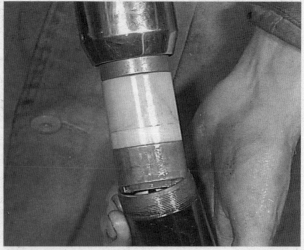

3.4a Removing the stanchion from the bottom leg (B44, B25 and C25 models)

3.4b Sliding off the inner sleeve and top bearing

3.5 Shuttle valve and end cap removed (earlier models)

3.6 Replace the bottom bush if badly worn

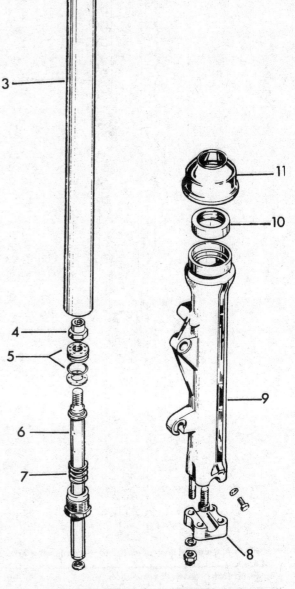

FIG. 4.6. EXPLODED VIEW OF THE SHUTTLE VALVE TYPE FORKS FITTED TO 1971 AND LATER MODELS

1 Top cap
2 Main spring
3 Fork stanchion
4 Shuttle valve retaining nut
5 Shuttle valve assembly
6 Damping rod
7 Recoil spring
8 Wheel spindle clamp
9 Bottom leg assembly
10 Stanchion seal
11 Rubber boot

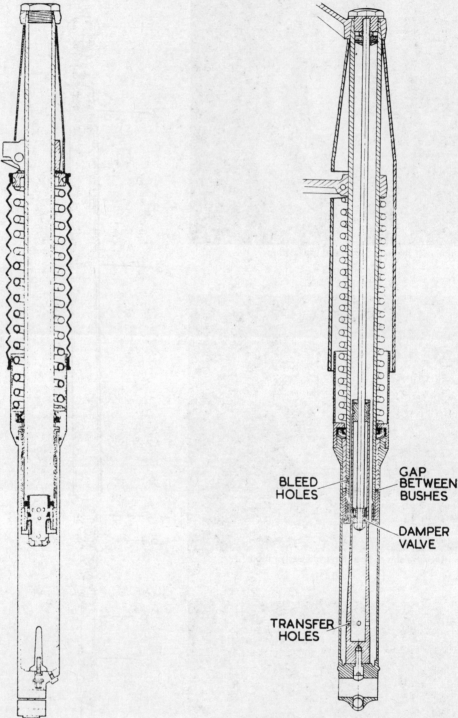

Fig. 4.7. Sectional view of earlier type shuttle valve fork leg

Fig. 4.8. Sectional view of the rod damper type fork leg

4 Front forks - general examination

1 Apart from the oil seals and bushes, it is unlikely that the forks will require any additional attention, unless the fork springs are weak or have to be replaced with stronger springs when the machine is to be used with a sidecar attached. If the fork legs or yokes have been damaged in an accident, it is preferable to have them replaced. Repairs are seldom practicable without the appropriate repair equipment and jigs, furthermore there is also the risk of fatigue failure.

2 Check the inner fork tubes (stanchions) for straightness by rolling them along a sheet of plate glass. If they are bowed to a greater extent than 5/32 inch, they will have to be renewed. If less than this tolerance an attempt to straighten them can be made using a surface plate and a soft-faced mallet.
3 Examine the top and bottom retaining yokes for cracks or distortion and repair or replace as necessary. Temporarily insert the fork stanchions into the top yoke tapers and check that both legs are parallel to each other both fore and aft and side to side.
4 Check both coil springs for cracks and measure their lengths. Both springs must be within ¼ inch of their manufactured dimensions, (see Specification at the beginning of this Chapter).

5 Front forks - examination and replacement of bushes and oil seals

1 If the fork legs have shown a tendency to leak oil or if there is any other reason to suspect the condition of the oil seals, now is the time to replace them. It is, in any case, rather pointless to replace old oil seals once they have been removed.
2 Some indication of the extent of wear of the fork bushes can be gained when the forks are being dismantled. Pull each fork inner tube out until it reaches the limit of its extension and check the side play. In this position the two fork bushes are closest together, which will show the amount of play to its maximum. Only a small amount of play that is just perceptible can be tolerated. If the play is greater than this, the bushes are due for replacement.
3 It is possible to check for play in the bushes whilst the forks are still attached to the machine. If the front wheel is gripped between the knees and the handlebars rocked to and fro, the amount of wear will be magnified by the leverage at the handle-bar ends. Cross-check by applying the front brake and pushing and pulling the front wheel backwards and forwards. It is important not to confuse any play that is evident with slackness in the steering head bearings, which should be taken up first.

6 Steering head bearings - examination and replacement

1 Before commencing to reassemble the forks, inspect the steering head races. The ball bearing tracks should be polished and free from indentations and cracks. If signs of wear are evident, the cones and cups must be replaced. They are a tight press fit and must be drifted out of position. A BSA Service Tool No. 61-3063 is available for extracting the cups that remain within the steering head assembly of the frame. It screws into the threaded centre of each cup and is driven out from the opposite end, bringing the cup with it.

2 Ball bearings are cheap. If there is any reason to suspect the condition of the existing ball bearings, they should be replaced without question. Note that each race is not completely full of ball bearings. Space should be left for the theoretical insertion of one extra ball, so that the race is not crowded, forcing the ball bearings to skid against one another. There are twenty ball bearings in each race, twenty-four in Star series models.
3 Use thick grease to retain the ball bearings in position, whilst the head stem is being assembled and adjusted.

7 Front forks - reassembly

1 To reassemble the forks, follow the dismantling procedure in reverse. Take particular care when passing the sliding fork members through the oil seals, which should be fitted with the lip facing downwards. It is a wise precaution to wind a turn or so of medium twine around the undercut at the base of the thread of the plated collars, to act as an extra seal.
2 Tighten the steering head carefully, so that all play is eliminated without placing undue stress on the bearings. The adjustment is correct if all play is eliminated and the handlebars will swing to full lock of their own accord when given a light push on one end.
3 It is possible to place several tons pressure quite unwittingly on the steering head bearings, if they are over-tightened. The usual symptom of over-tight bearings is a tendency for the machine to roll at low speeds, even though the handlebars may appear to turn quite freely.
4 One problem that will arise during reassembly is the reluctance of the main stanchions to pass up into the fork top yoke. BSA service tool No. 61-3350 is used for this purpose; it threads into the top of each fork tube and can be used to pull the stanchion upward so that the tapered end engages with the fork yoke. If the tool is not available, a broom handle of the correct diameter can be used with equal effect, if it is first screwed into the end of the thread of each stanchion. Care should be taken in this instance, to prevent particles of wood from falling into the fork tubes.

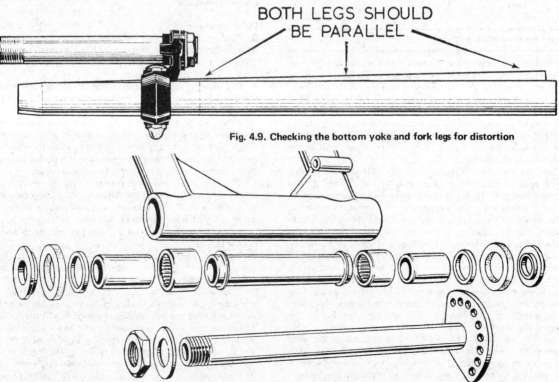

Fig. 4.9. Checking the bottom yoke and fork legs for distortion

Fig. 4.10. Exploded view of needle roller swinging arm pivot bearing, fitted to 1971 and later models

5 If, after assembly, it is found that the forks are incorrectly aligned or unduly stiff in action, loosen the front wheel spindle, the two caps at the top of the fork legs and the pinch bolts in both the top and bottom yokes. The forks should then be pumped up and down several times to realign them. Retighten all the nuts and bolts in the same order, finishing with the steering head pinch bolt.

6 This same procedure can be used if the forks are misaligned after an accident. Often the legs will twist within the fork yokes, giving the impression of more serious damage, even though no structural damage has occurred.

7 Do not omit to add the correct amount of damping oil to each fork leg before replacing the fork leg caps. See specifications list for the amount and viscosity of oil to be added.

7.7 Top up both fork legs to the correct level (see specifications)

8 Front forks - damping action

1 Each fork leg contains a predetermined quantity of oil of recommended viscosity, which is used as a damping medium to control the action of the compression springs within the forks when various road shocks are encountered. If the damping fluid is absent, there is no control over the rebound action of the fork springs and fork movement will be excessive, giving a very 'lively' ride. Damping restricts fork movement on the rebound and is progressive in action - the effect becomes more powerful as the rate of deflection increases.

2 When the rod damper type forks are deflected, a double damping action occurs within each fork leg. As the coil springs are compressed and the lower leg rises the oil within the damper tube is compressed by the damper rod and forced through the transfer holes into the bottom section of the main leg. Simultaneously the oil already in the main leg is forced upwards between the outside of the damper tube and top tube bush. The damping tube is tapered towards the bottom end of the leg and the pressure of the oil increases as the damping rod approaches the bottom of the tube, providing a progressively firmer damping action. When oil begins to flow into the top tube it passes into an area between the fork leg and the top tube bushes. Maximum fork deflection is cushioned by the oil that remains in the bottom of the lower leg.

3 When the fork begins to extend, oil is compressed between the two bushes and forced through the four bleed holes back into the lower leg reservoir. The rising damper valve creates a partial vacuum in the tube and draws oil through the transfer holes in the damping tube thus smoothing out the spring recoil action.

4 The shuttle valve type forks fitted to later models operate in a similar manner with the exception that the transfer of oil through the bleed holes in the fork stanchion is achieved by the movement of the sliding shuttle valve. Before the main stanchion can reach its limit of travel, a tapered plug in the bottom of the lower fork leg enters the hole in the bottom of the shuttle valve and slows down the rate of oil transfer until it is virtually cut off altogether. The remaining oil is incompressible and the fork leg is therefore prevented from 'bottoming'.

9 Frame assembly - examination and renovation

1 The frame on all unit construction models is exceptional robust and should not require attention unless it is damaged in an accident. Frame repairs are best entrusted to a specialist in this type of repair work, who will have all the necessary jigs and mandrels available to ensure correct alignment. In many instances, a replacement frame from a breaker's yard is the cheaper and more satisfactory alternative.

2 If the machine is stripped for an overhaul, this affords an excellent opportunity to inspect the frame for signs of cracks or other damage that may have occurred in service. Check the front down tube at the point immediately below the steering head, which is where the break is most likely to occcur. Check the top tube of the frame for straightness - this is the tube most likely to bend in the event of an accident.

10 Swinging arm rear suspension - examination and renovation

1 After an extended period of service, the bush and pivot pin of the swinging arm fork will wear, giving rise to lateral play that will affect the handling characteristics of the machine.

2 Three types of bearings were used on BSA single cylinder machines: plain bushes, rubber 'silentbloc' brushes and, on the later machines, needle roller bearings. The dismantling and renovation procedure is given under separate headings in the following sections.

Plain bush type

3 First remove the rear wheel and chain guard, (refer to the appropriate section in this manual for the correct procedure). Take off the dual seat and remove the coil, rectifier and horn from the subframe crosspiece.

4 Disconnect the rear light cable at the snap connector and remove the mudguard attachment bolts to withdraw the mudguard.

5 After the two rear shock absorbers have been removed the complete rear subframe can be detached after unscrewing the single bolt adjacent to the rear of the petrol tank and a bolt at each end of the swinging arm spindle. If a grease nipple is fitted to the spindle housing remove it to avoid damage.

6 The pivot spindle will probably be extremely tight and ideally a tool as shown in the adjacent illustration should be used to withdraw it. If this type of extractor cannot be borrowed or made up, a hammer and drift will have to be used, but extreme care should be taken not to damage the spindle threads or even buckle the spindle.

7 Remove the swinging arm assembly and drive out the bushes using a suitable drift. Insert the new bushes with the split facing towards the front of the machine.

8 If reaming facilities are not available at home take the swinging arm assembly and spindle to a garage or engineering works and have the bearings line reamed to suit the spindle.

9 Grease the bushes and spindle well and reassemble onto the machine, ensure the spindle shoulder is flush with the swinging arm bosses on both sides before replacing the thrust washers and tightening the nuts: replace the nipple and apply the grease gun.

10 Reassemble the subframe, mudguard, wheel etc., in the reverse order of dismantling, cleaning off any accumulated road dirt and grease where necessary. Ensure that the rear brake and

chain are correctly adjusted before trying the bike out on the road.

Rubber 'Silentbloc' bush type

11 Remove the rear wheel, chainguard, shock absorbers, and rear brake pedal and disconnect the brake light switch holding the bracket to the frame plate. The brake pedal stop is held by one nut and must also be removed.

12 Remove the large pivot nut and washer on the right side of the machine after first removing the footrest support plate and drive the shaft out of the swing arm bore using a soft metal drift.

13 Tap the left side of the swing arm down and the right side up, using a mallet. This will free the swing arm from the frame plates.

14 Each swing arm 'silentbloc' bush consists of two steel sleeves bonded together with rubber. The inner sleeves are slightly longer that half the width of the swing arm and are locked together, thereby putting the rubber under tension when the arm swings through its arc.

If it is necessary to replace the bushes, the rubber must first be burned out to facilitate removal. This can be done with a thin rod or strip of metal heated until red hot.

15 When enough rubber has been removed, drive out the inner sleeves and then the outer sleeves.

16 Fit new rubber bushes making sure that they are free from grease or oil and reassemble the various component in the reverse order of removal. Do not tighten the pivot shaft until after the shock absorbers have been correctly fitted.

Needle bearing type

17 Remove the rear wheel assembly and remove the brake pedal to provide clearance for taking out the swing arm shaft, if necessary.

18 Remove the shock absorber bottom mounting bolts and swing the units out of the way.

19 Unscrew the pivot shaft nut and drive the shaft out, using a suitable drift.

20 The outer oil seals and spacers can be pulled out by hand, and the inner bushings can be withdrawn using long nosed pliers. Take care not to damage the bearing surface of the bushes if they are to be replaced. The bearings can then be driven out using a suitable drift. When installing the bearings, use a drift of the correct size to avoid damaging them, and install the bearings squarely in the housing. Assemble the pivot components using the accompanying illustration as a guide, grease all the bearing components well before reassembly.

10.11 Remove brake stop if fitted in place of bracket bolt

10.12a Remove footrest bracket ...

10.12b ... and then the pivot shaft nut

10.12c Withdrawing the pivot shaft

10.13 Removing the complete swinging arm assembly

10.14 Replace 'Silent bloc' bushes if worn

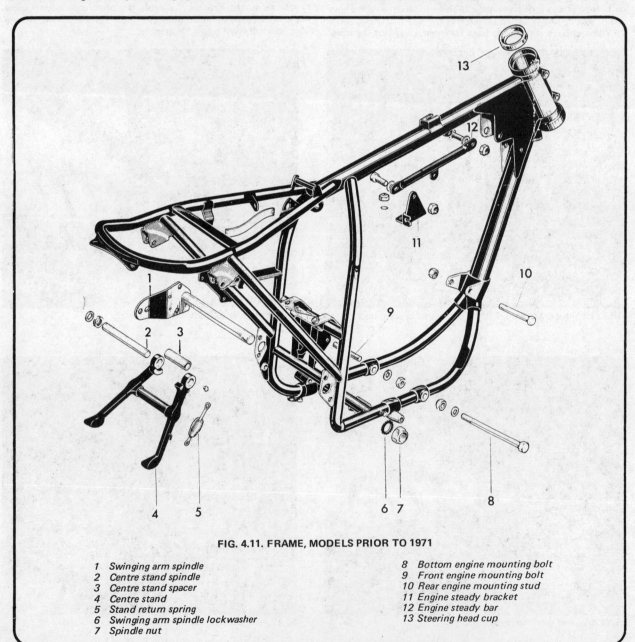

FIG. 4.11. FRAME, MODELS PRIOR TO 1971

1 Swinging arm spindle	8 Bottom engine mounting bolt
2 Centre stand spindle	9 Front engine mounting bolt
3 Centre stand spacer	10 Rear engine mounting stud
4 Centre stand	11 Engine steady bracket
5 Stand return spring	12 Engine steady bar
6 Swinging arm spindle lockwasher	13 Steering head cup
7 Spindle nut	

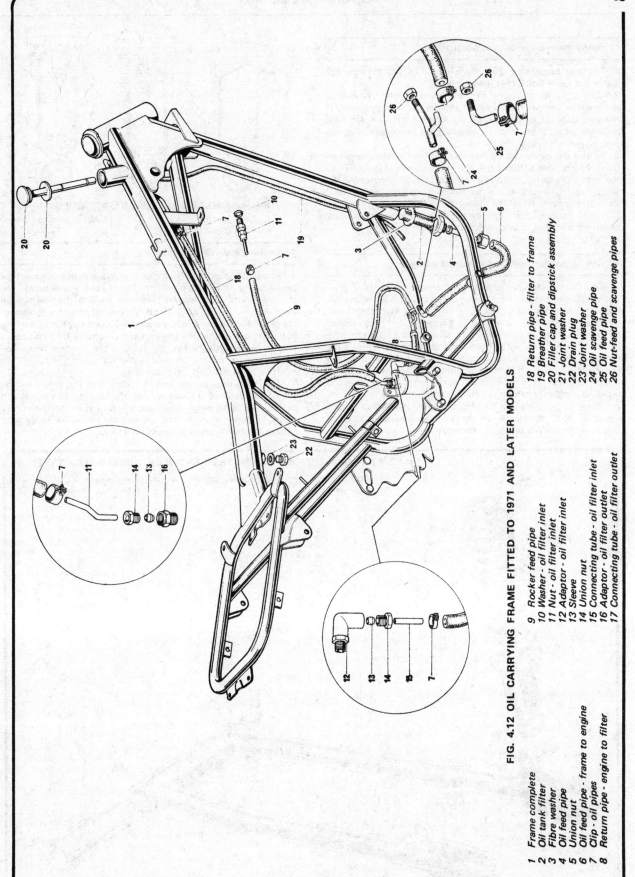

FIG. 4.12 OIL CARRYING FRAME FITTED TO 1971 AND LATER MODELS

1 Frame complete
2 Oil tank filter
3 Fibre washer
4 Oil feed pipe
5 Union nut
6 Oil feed pipe - frame to engine
7 Clip - oil pipes
8 Return pipe - engine to filter

9   Rocker feed pipe
10 Washer - oil filter inlet
11 Nut - oil filter inlet
12 Adaptor - oil filter inlet
13 Sleeve
14 Union nut
15 Connecting tube - oil filter inlet
16 Adaptor - oil filter outlet
17 Connecting tube - oil filter outlet

18 Return pipe - filter to frame
19 Breather pipe
20 Filler cap and dipstick assembly
21 Joint washer
22 Drain plug
23 Joint washer
24 Oil scavenge pipe
25 Oil feed pipe
26 Nut-feed and scavenge pipes

11 Rear suspension units - examination

1   The rear suspension units are removed by withdrawing the upper and lower bolts, nuts, and washers.

2   Only a limited amount of dismantling can be undertaken because the damper unit is sealed and cannot be dismantled. If the unit leaks oil, or if the damping action is lost, the unit must be replaced as a whole after removing the compression spring and, on earlier models, the shroud.

3   Before the spring can be removed, the unit must be detached from the machine and clamped in a vice. If pressure is applied to the top of the spring, (shroud on earlier models) compressing the spring, the split collets can be removed and the spring (and shroud) released. Note the spring is colour-coded; the colour relates to the spring rating. Springs can be obtained in a variety of different ratings, to accommodate different loadings.

12 Rear suspension units - adjusting the setting

1   The Girling rear suspension units fitted to the later BSA single cylinder models have a three-position cam adjuster built into the lower portion of the leg to suit varying load conditions, on later models the cam ring is concealed by a castellated adjusting nut which operates in the same sense as the cam ring fitted to earlier machines. The lowest position should suit the average rider, under normal road conditions. When a pillion passenger is carried, the second or middle position offers a better choice and for continuous high speed work or off-the-road riding, the highest position is recommended.

2   These adjustments can be effected without need to detach the units. A 'C' spanner in the tool kit is used to rotate the cam ring until the desired setting is obtained.

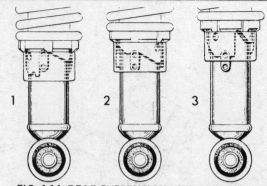

FIG. 4.14. REAR SUSPENSION UNIT ADJUSTMENT
1   Light setting   3   Heavy setting
2   Medium setting

13 Centre stand - examination

1   Some BSA unit construction models are provided with a centre stand attached to lugs on the bottom frame tubes. The stand provides a convenient means of parking the machine on level ground, or for raising one or other of the wheels clear of the ground in the event of a puncture. The stand pivots on a long bolt that passes through the lugs and is secured by a nut and a washer. A return spring return retracts the stand so that when the machine is pushed forward it will spring up and permit the machine to be wheeled, prior to riding.

2   The condition of the return spring and the return action should be checked frequently, also the security of the nut and bolt. If the stand drops whilst the machine is in motion, it may catch in some obstacle and unseat the rider. On some later models the centre stand is an optional extra.

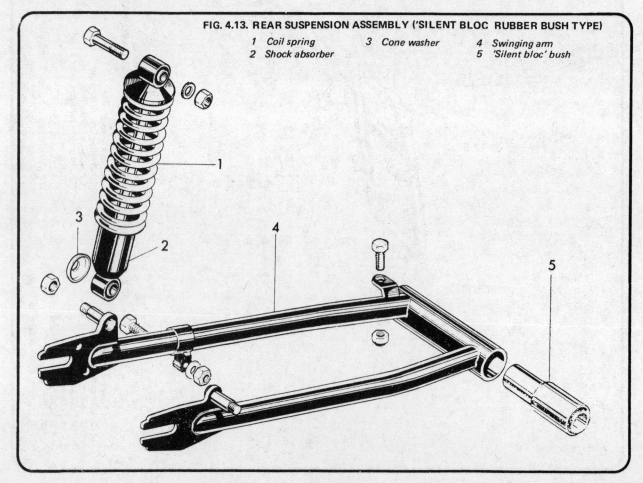

FIG. 4.13. REAR SUSPENSION ASSEMBLY ('SILENT BLOC RUBBER BUSH TYPE)
1   Coil spring   3   Cone washer   4   Swinging arm
2   Shock absorber   5   'Silent bloc' bush

14 Prop stand - examination

1 A prop stand that pivots from a lug at the front end of the lower left-hand frame tube provides an additional means of parking the machine. This too has a return spring, which should be strong enough to cause the stand to retract immediately the machine is raised into a vertical position. It is important that this spring is examined at regular intervals, also the nut and bolt that act as the pivot. A falling prop stand can have far more serious consequences if it should fall whilst the machine is on the move.

15 Footrests - examination and renovation

1 The footrests, which bolt to the frame lugs, are malleable and will bend if the machine is dropped. Before they can be straightened, they must be detached from the frame and have the rubbers removed.
2 To straighten the footrests, clamp them in a vice and apply leverage from a long tube that slips over the end. The area in which the bend has occurred should be heated to a cherry red with a blow lamp, during the bending operation. Do not bend the footrests cold, otherwise there is risk of a sudden fracture.

16 Speedometer - removal and replacement

1 The BSA unit-construction models are fitted with a Smiths Chronometric speedometer, calibrated up to 120 mph, (80 mph on smaller machines). An internal lamp is provided for illuminating the dial during the hours of darkness and the odometer has a trip setting, so that the lower mileage reading can be set to zero before a run is commenced.
2 The speedometer head has two studs, which permit it to be attached to either a bracket on the fork top yoke or to a nacelle or cowl by means of a simple clamp arrangement. On the earlier models, the speedometer head is mounted in the top of the headlamp cowl, where it is retained by a clamp and a strainer bolt. First remove the headlight from the cowl and then remove the retaining nut(s) to the cowl, after the drive cable has been detached.
3 Apart from defects in the drive or the drive cable itself, a speedometer that malfunctions is difficult to repair. Fit a replacement or alternatively entrust the repair to an instrument repair specialist, bearing in mind that the speedometer must function in a satisfactory manner to meet statutory requirements.
4 If the odometer readings continue to show an increase, without the speedometer indicating the road speed, it can be assumed the drive and drive cable are working correctly and that the speedometer head itself is at fault.

17 Speedometer cable - examination and renovation

1 It is advisable to detach the speedometer drive cable from time to time in order to check whether it is adequately lubricated, and whether the outer covering is compressed or damaged at any point along its run. A jerky or sluggish speedometer movement can often be attributed to a cable fault.
2 To grease the cable, withdraw the inner cable. After removing the old grease, clean with a petrol soaked rag and examine the cable for broken strands or other damage.
3 Regrease the cable with high melting point grease and ensure that there is no grease on the last six inches, at the end where the cable enters the speedometer head. If this precaution is not observed, grease will work into the speedometer head and immobilise the movement.
4 Inspection will show whether the speedometer drive cable has broken. If so, the inner cable can be removed and replaced with

another whilst leaving the outer cable in place - provided the outer cable is not damaged or compressed at any point along its run. Measure the cable length exactly when purchasing a replacement, because this measure is critical.

18 Tachometer - removal and replacement

1 The tachometer drive when fitted, is normally taken from one of the timing pinions, through an aperture in the left hand crankcase half which will accept the drive gearbox. The tachometer head may be of either the Chronometric or magnetic type, depending on the year of manufacture.
2 It is not possible to effect a satisfactory repair to a defective tachometer head, hence replacement is necessary if the existing head malfunctions. Make sure an exact replacement is obtained; some tachometer heads work at half-speed if a different type of drive gearbox is employed.
3 The tachometer head is illuminated internally so that the dial can be read during the hours of darkness.

19 Tachometer drive cable - examination and renovation

1 Although a little shorter in length, the tachometer drive cable is identical in construction to that used for the speedometer drive. The advice given in Section 17 of this Chapter applies also the the tachometer drive cable.

20 Tachometer drive gearbox - examination

1 The tachometer drive gearbox is unlikely to give trouble during the normal service life of the machine, provided it is greased regularly at the external grease nipple.
2 If the drive gearbox has to be replaced, make sure it is replaced with one having an identical drive ratio, otherwise the tachometer head readings will no longer be true.
3 There should be a good joint between the flange of the drive gearbox and the flange of the timing cover if oil leaks are to be prevented. Always use a new gasket and preferably a little gasket cement if the joint has to be broken and re-made.

21 Dualseat - removal

1 The dualseat is attached by means of two bolts towards the rear end of the underpan, one on each side of the machine. If these bolts are removed, the dualseat will lift off, after the slotted end of the nose connection has been pulled clear from the mounting tube across the frame.

22 Tank badges and motifs

1 Different types of tank badge have been used since the inception of the BSA twin cylinder models, ranging from the BSA winged motif to circular or pear shaped plastics badges that either screw direct to the petrol tank or are retained by small hidden brackets.
2 The plastics badges are mounted over a thin rubber mat, to eliminate vibration chatter and to give some form of cushioning when the centre retaining screw is tightened. Certain colour schemes apply to certain models and if a badge is lost it is important to specify the colour scheme when endeavouring to obtain a replacement.

23 Steering head lock

1 Some models are fitted with a steering head lock inserted into the fork top yoke. If the forks are turned to the extreme left, they can be locked in this position to prevent theft. The

lock is of Yale manufacture.

2   Add an occasional few drops of thin machine oil to keep the lock in good working order. This should be added to the periphery of the moving drum and NOT the keyhole.

24 Steering damper - use

1   Mention has been made of the steering damper fitted to some machines and comprising of a friction plate device that can be adjusted to vary the amount of effort required to turn the handlebars. In many respects, the steering damper can be regarded as a legacy of the past when it was necessary to counteract the tendency of some machines to develop a speed wobble at high speeds. Today, the steering damper comes into its own mainly when a sidecar is attached, since it will prevent the handlebars from oscillating at very low speeds, when it is applied.

2   Under normal riding conditions, the steering damper can be slackened off. It is sometimes advantageous to have it biting just a trifle at high speeds, to ease the strain on the arms.

3   To remove the steering damper, detach the split pin at the extreme end of the damper rod and/or unscrew the knob at the top of the steering head until it can be drawn away with the rod attached. When the fixed plate is detached from below the bottom yoke of the forks, the friction plates can be removed for inspection. They seldom require attention.

25 Cleaning - general

1   After removing all surface dirt with a rag or sponge that is washed frequently in clean water, the application of car polish or wax will restore a good finish to the cycle parts of the machine after they have dried thoroughly. The plated parts should require only a wipe with a damp rag, although it is permissible to use a chrome cleaner if the plated surfaces are badly tarnished.

2   Oil and grease, particularly when they are caked on, are best removed with a proprietary cleanser such as 'Gunk' or 'Jizer'. A few minutes should be allowed for the cleanser to penetrate the film of oil and grease before the parts concerned are hosed down. Take care to protect the magneto, carburettor and electrical parts from the water, which may cause them to malfunction.

3   Polished aluminium alloy surfaces can be restored by the application of Solvol 'Autosol' or some similar polishing compound, and the use of a clean duster to give the final polish.

4   If possible, the machine should be wiped over immediately after it has been used in the wet, so that it is not garaged under damp conditions that will promote rusting. Make sure to wipe the chain and if necessary re-oil it, to prevent water from entering the rollers and causing harshness with an accompanying high rate of wear. Remember there is a little chance of water entering the control cables if they are lubricated regularly, as recommended in the Routine Maintenance Section.

26 FAULT DIAGNOSIS — FRAME AND FORKS

Symptom	Reason/s	Remedy
Machine is unduly sensitive to road conditions	Forks and/or rear suspension units have defective damping	Check oil level in forks. Replace rear suspension units.
Machine tends to roll at low speeds	Steering head bearings overtight or damaged	Slacken bearing adjustment. If no improvement, dismantle and inspect bearings.
Machine tends to wander, steering is imprecise	Worn swinging arm bearings	Check and if necessary renew bearings.
Fork action stiff	Fork legs have twisted in yokes or have been drawn together at lower ends	Slacken off spindle nut clamps, pinch bolts in fork yokes and fork top nuts. Pump forks several times before retightening from bottom. Is distance piece missing from fork spindle?
Forks judder when front brake is applied	Worn fork bushes Steering head bearings too slack	Strip forks and replace bushes. Readjust, to take up play.
Wheels out of alignment	Frame distorted as result of accident damage	Check frame alignment after stripping out. If bent, specialist repair is necessary.

Chapter 5 Wheels, brakes and tyres

Contents

Specifications

Wheel sizes

Front	20 inch diameter - B25T, B50T and B50MX models	
	18 inch diameter - all other models	
Rear	18 inch - all models	

Tyre sizes

Front	3.00 inch x 20 inch B25T, B50T and B50MX	
... ...	3.25 inch x 18 inch all other models	
Rear	3.50 inch x 18 inch - all models 1958 onwards	

Tyre pressures

Front	22 psi	
Rear	24 psi*	

* increase to 28 psi when a pillion passenger is carried.

Brake dimensions

Front	6 inch diameter, 1½ inch width - all models 1958 - 1966	
	8 inch diameter, 15/8 inch width - 1968 B44 models only	
	7 inch diameter, 1½ inch width - all models 1966 - 1970	
	6 inch diameter, .875 inch width - B25SST, B50T and B50MX only	
	8 inch diameter, 1½ inch width - 1971 B25 and B50SS models	
Rear	7 inch diameter, 11/8 inch width - all models	

1 General description

The first BSA single cylinder, unit-construction models comprising the 'Star' series introduced in 1958, were equipped with a 6" full width hub front brake and a 7" rear brake. In 1966 the Shooting Star and Starfire models were introduced with a 7" half hub front 'stopper'. The 7" rear brake was retained on these and all later models.

From 1969 until 1970 all BSA singles were fitted with a 7" twin leading shoe front brake housed in a full width hub. For the 1971 season models were offered with either a 6" single leading shoes front brake or an 8" double leading shoe type, depending on specification. The latter type of front brake is extremely powerful and newcomers to a machine thus equipped should exercise caution until experience is gained.

All 1966 to 1970 models featured a quickly detachable rear wheel enabling it to be easily removed without disturbing the brake assembly or chain.

2 Front wheel - examination and renovation

1 Place the machine on the centre stand so that the front wheel is raised clear of the ground. Spin the wheel and check the rim for alignment. Small irregularities can be corrected by tightening the spokes in the affected area, although a certain amount of experience is necessary if over-correction is to be avoided. Any 'flats' in the wheel rim should be evident at the same time. These are more difficult to remove with any success and in most cases the wheel will have to be rebuilt on a new rim. Apart from the effect on stability, especially at high speeds, there is much greater risk of damage to the tyre beads and walls if the machine is ridden with a deformed wheel.

2 Check for loose or broken spokes. Tapping the spokes is the best guide to tension. A loose spoke will produce a quite different note that should be tightened by turning the nipple in an anti-clockwise direction. Always re-check for run-out by spinning the wheel again.

3.1a Removal of the retaining nut ...

3 Front brake assembly - examination, renovation and re-assembly

1 For details on front wheel removal refer to Chapter 4, Section 2. On the 1966-1967 models fitted with the quickly detachable front spindle the brake plate is just a push fit and can be lifted straight off. On all other models the right-hand retaining nut must be removed first. If the plate appears to jam move the brake operating lever in an 'off' direction to relieve any friction between linings and drum.

2 Before dismantling the brake assembly, examine the condition of the brake linings. If they are wearing thin or unevenly, they must be replaced.

3 To remove the brake shoes from the brake plate, position the operating cam(s) so that the shoes are in the fully-expanded position and pull them apart whilst lifting them upwards, in the form of a 'V'. When they are clear of the brake plate the return springs should be removed and the shoes separated.

4 It is possible to replace the brake linings fitted with rivets and not bonded on, as is the current practice. Much will depend on the availability of the original type of linings; service-exchange brake shoes with bonded-on linings may be the only practical form of replacement.

5 If new linings are fitted by rivetting, it is important that the rivet heads are countersunk, otherwise they will rub on the brake drum and be dangerous. Make sure the lining is in very close contact with the brake shoes during the rivetting operation; a small 'C' clamp of the type used by carpenters can often be used to good effect until all the rivets are in postiion. Finish off by chamfering off the end of each shoe, otherwise fierce brake grab may occur due to the pick-up of the leading edge of each lining.

6 Before replacing the brake shoes, check that the brake shoes, check that the brake operating cam(s) is/are working smoothly and not binding in its pivot. The cam(s) can be removed for greasing by unscrewing the nut on the end of the brake operating arm and drawing the arm off so that the cam(s) and spindle can be withdrawn from the inside of the brake plate.

7 Check the inner surface of the brake drum, on which the brake shoes bear. The surface should be smooth and free from indentations, or reduced braking efficiency is inevitable. Remove all traces of brake lining dust and wipe the surface with a rag soaked in petrol to remove any traces of grease or oil.

8 To reassemble the brake shoes on the brake plate, fit the return springs and force the shoes apart, holding them in a 'V' formation. If they are now located with the brake operating arm and pivot, they can usually be snapped into position by pressing downward. Do not use excessive force, or the shoes may be distorted permanently.

3.1b ... allows the back plate to be lifted out complete with brake shoes

3.3 The shoe can be removed by lifting upwards and then outwards over the cam(s)

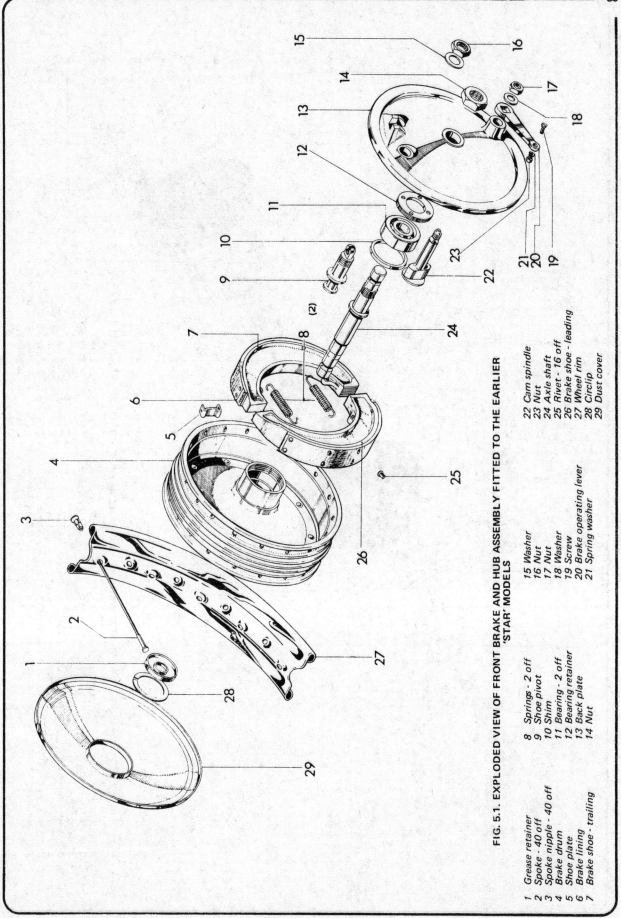

FIG. 5.1. EXPLODED VIEW OF FRONT BRAKE AND HUB ASSEMBLY FITTED TO THE EARLIER 'STAR' MODELS

1 Grease retainer
2 Spoke - 40 off
3 Spoke nipple - 40 off
4 Brake drum
5 Shoe plate
6 Brake lining
7 Brake shoe - trailing

8 Springs - 2 off
9 Shoe pivot
10 Shim
11 Bearing - 2 off
12 Bearing retainer
13 Back plate
14 Nut

15 Washer
16 Nut
17 Nut
18 Washer
19 Screw
20 Brake operating lever
21 Spring washer

22 Cam spindle
23 Nut
24 Axle shaft
25 Rivet - 16 off
26 Brake shoe - leading
27 Wheel rim
28 Circlip
29 Dust cover

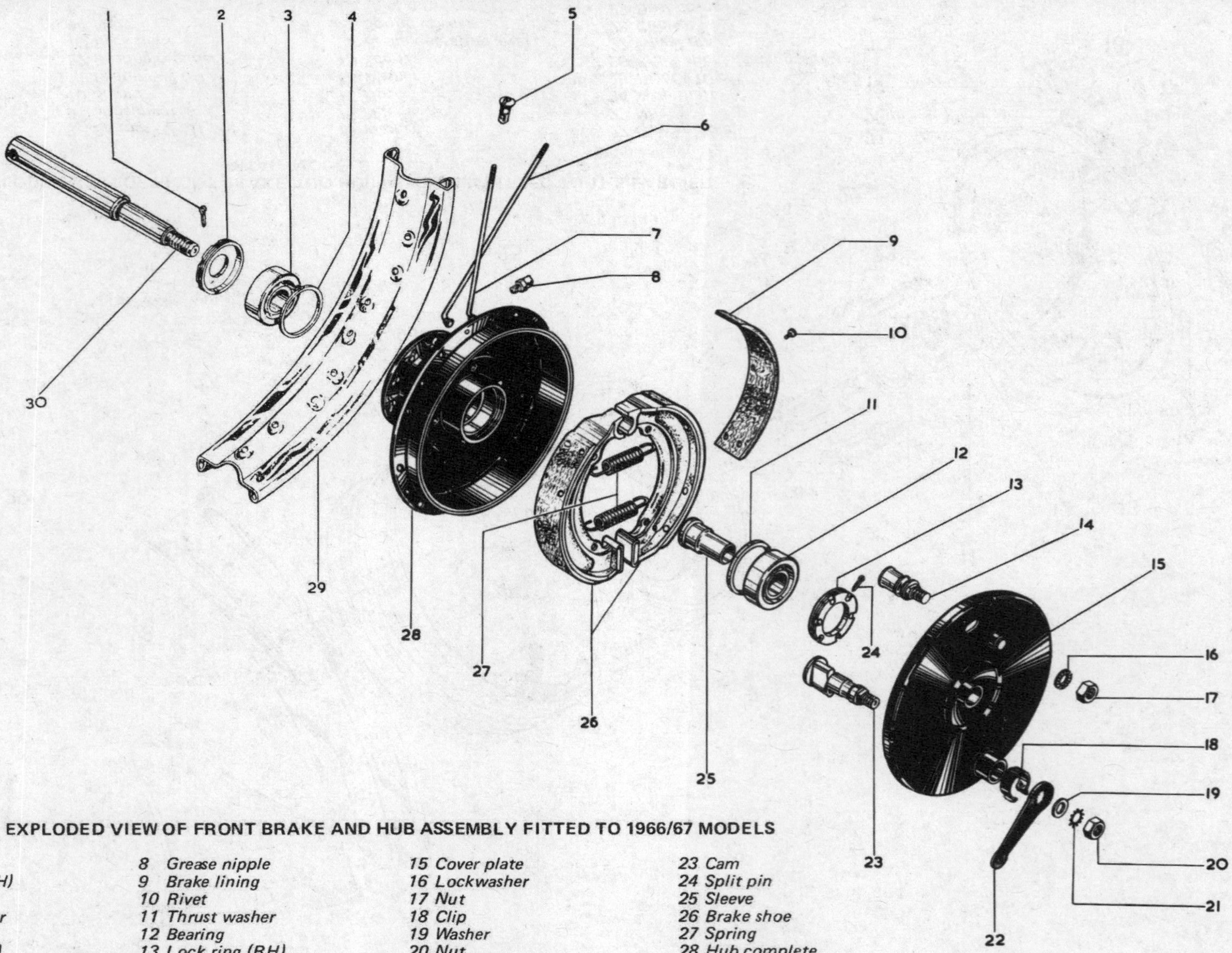

FIG. 5.2. EXPLODED VIEW OF FRONT BRAKE AND HUB ASSEMBLY FITTED TO 1966/67 MODELS

1 Split pin
2 Lock ring (LH)
3 Bearing
4 Thrust washer
5 Nipple
6 Spoke (short)
7 Spoke (long)
8 Grease nipple
9 Brake lining
10 Rivet
11 Thrust washer
12 Bearing
13 Lock ring (RH)
14 Fulcrum pin
15 Cover plate
16 Lockwasher
17 Nut
18 Clip
19 Washer
20 Nut
21 Lockwasher
22 Brake lever
23 Cam
24 Split pin
25 Sleeve
26 Brake shoe
27 Spring
28 Hub complete
29 Rim (WM2-18)
30 Spindle

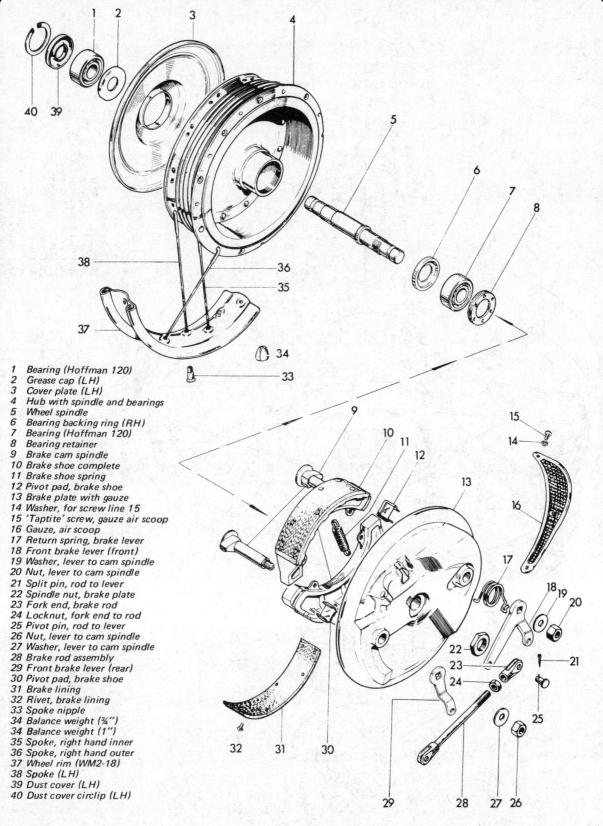

1   Bearing (Hoffman 120)
2   Grease cap (LH)
3   Cover plate (LH)
4   Hub with spindle and bearings
5   Wheel spindle
6   Bearing backing ring (RH)
7   Bearing (Hoffman 120)
8   Bearing retainer
9   Brake cam spindle
10 Brake shoe complete
11 Brake shoe spring
12 Pivot pad, brake shoe
13 Brake plate with gauze
14 Washer, for screw line 15
15 'Taptite' screw, gauze air scoop
16 Gauze, air scoop
17 Return spring, brake lever
18 Front brake lever (front)
19 Washer, lever to cam spindle
20 Nut, lever to cam spindle
21 Split pin, rod to lever
22 Spindle nut, brake plate
23 Fork end, brake rod
24 Locknut, fork end to rod
25 Pivot pin, rod to lever
26 Nut, lever to cam spindle
27 Washer, lever to cam spindle
28 Brake rod assembly
29 Front brake lever (rear)
30 Pivot pad, brake shoe
31 Brake lining
32 Rivet, brake lining
33 Spoke nipple
34 Balance weight (¾")
34 Balance weight (1")
35 Spoke, right hand inner
36 Spoke, right hand outer
37 Wheel rim (WM2-18)
38 Spoke (LH)
39 Dust cover (LH)
40 Dust cover circlip (LH)

FIG. 5.3. EXPLODED VIEW OF FRONT BRAKE AND HUB ASSEMBLY FITTED TO
1968 TO 1970 MODELS

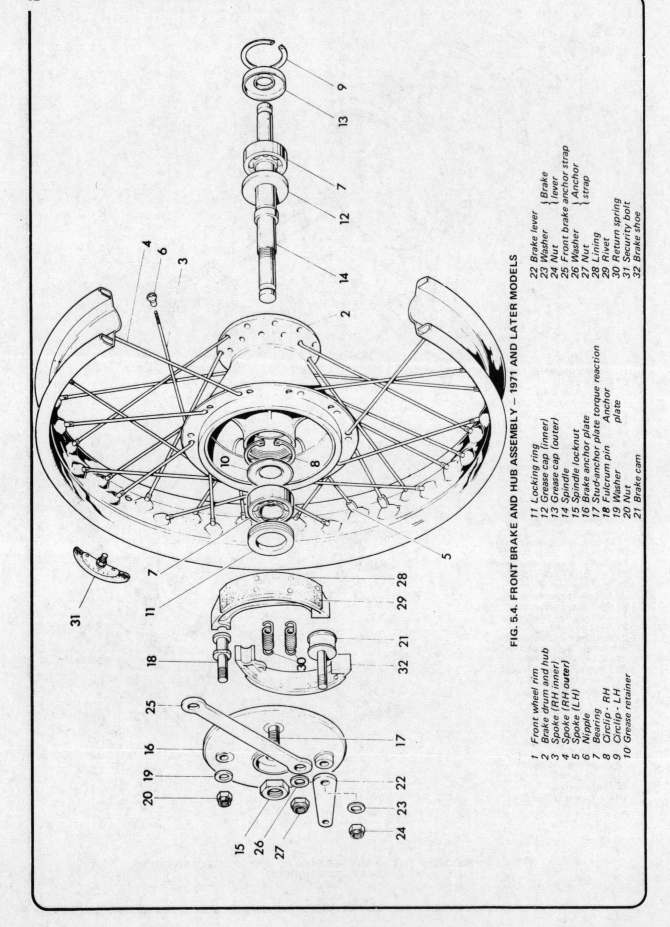

FIG. 5.4. FRONT BRAKE AND HUB ASSEMBLY – 1971 AND LATER MODELS

1 Front wheel rim
2 Brake drum and hub
3 Spoke (RH inner)
4 Spoke (RH outer)
5 Spoke (LH)
6 Nipple
7 Bearing
8 Circlip - RH
9 Circlip - LH
10 Grease retainer
11 Locking ring
12 Grease cap (inner)
13 Grease cap (outer)
14 Spindle
15 Spindle locknut
16 Brake anchor plate
17 Stud-anchor plate torque reaction
18 Fulcrum pin
19 Washer
20 Nut
21 Brake cam

16 Brake anchor plate
17 Stud-anchor plate torque reaction
18 Fulcrum pin   Anchor
19 Washer   plate
20 Nut
21 Brake cam

22 Brake lever
23 Washer   } Brake
24 Nut   } lever
25 Front brake anchor strap
26 Washer   } Anchor
27 Nut   } strap
28 Lining
29 Rivet
30 Return spring
31 Security bolt
32 Brake shoe

5.5 Apply grease sparingly to the brake lever pivot(s)

cannot be overstressed because the anchorage of the front brake is dependant on the correct location of these parts.

2 Some models have an alternative torque arm arrangement. In this case the torque arm must be slipped over the end of the stud in the brake plate before the brake spindle is inserted and tightened.

3 Replace the front wheel spindle, if of the detachable type, or replace the split clamps at the extreme end of each fork leg. Before tightening the pinch bolt that secures the front wheel spindle, depress the forks several times so that the left-hand fork leg can position itself correctly on the distance bush. (detachable spindle models only),

4 Reconnect the front brake cable and check that the brake functions correctly. If the connection through the brake operating arm is made by means of a clevis pin, a new split pin must be used to secure the clevis pin in postion. Re-check the wheel spindle nut, pinch bolt or the fork end bolts for tightness, also the nuts that secure the separate torque arm (if fitted).

5 On models fitted with grease nipples on the brake operating arm(s) a grease gun should be applied to ensure lubrication of the cam spindles(s). Do not pump in too much grease (one or two strokes of the gun should suffice) otherwise grease may find its way onto the brake linings and seriously impair braking efficiency.

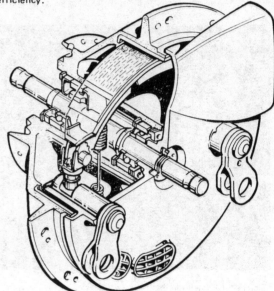

Fig. 5.5. 8" Twin leading shoe front brake assembly fitted to 1971 and later models

4 Wheel bearings - removal, examination and replacement

1 When the brake plate complete with brake assembly has been removed, the bearing retainer within the brake drum will be revealed. This has a RH thread on 1970 and earlier models and a LH thread on 1971 and later models. On half width hub brakes the retainer is secured with a split pin which must be removed. Unscrew the bearing retainer using either a peg spanner, (BSA service tool 61-3694) or a centre punch. Drift the bearing out of the hub after removal of the retainer by striking the LH end of the wheel spindle with a rawhide mallet. A grease retainer is inserted between the bearing and the shoulder of the spindle on some models and must not be omitted during reassembly; on half width hub models a bearing sleeve is a sliding fit on the spindle and must also be replaced.

2 The LH bearing can be removed by detaching the retaining circlip or, on some earlier models a second screwed retainer and drifting the bearing out from the RH side of the hub. On some models the bearing will be preceded by a dust cover.

3 Remove all the old grease from the hub and bearings, giving the latter a final wash in petrol. Check the bearings for play or signs of roughness when they are turned. If there is any doubt about their condition, play safe and renew them. A new bearing has no discernable play.

4 Before replacing the bearings, first pack the hub with new, high melting point grease. Then grease both bearings and drive them back into position, not forgetting any distance piece, hollow sleeve or shim washers that were fitted originally. Make sure the bearing retainer is tight and that the dust cover is located correctly. The bearing retainer performs the dual role of preventing grease from entering the brake drum, thereby reducing braking efficiency.

5 There is no means of adjusting wheel bearings of the ball journal type. If play is evident, the bearings have reached the end of their useful service life.

5 Front wheel - reassembly and replacement

1 Replace the front brake assembly in the brake drum and align the front wheel so that the projection on the brake plate engages with the peg on the right-hand fork leg. The importance of this

6 Rear wheel - examination, removal and renovation

1 Before removing the rear wheel, check for rim alignment, damage to the rim and loose or broken spokes by following the procedure that applies to the front wheel examination, in the preceding Section.

2 To remove the rear wheel on the earlier 'Star' models first disconnect the spring link and unwind the chain from the rear sprocket by slowly turning the wheel. Do not unwind the chain off the gearbox sprocket but let it hang down onto a piece of clean cloth.

3 Unscrew the knurled adjuster from the end of the brake rod and, if fitted, remove the speedometer cable from the drive unit. Unfasten the nuts securing the backplate anti-torque link and remove link completely.

4 Slacken the wheel spindle nuts but do not alter the setting of the chain adjusters - this will avoid the necessity of re-adjusting the chain when the wheel is replaced. The wheel can now be withdrawn from the frame, to facilitate removal it may be necessary to place a box or thick block of wood beneath the

centre stand to raise the rear of the machine.

5   The procedure for rear wheel removal on 1971 and later machines is virtually identical to that described for the 'Star' series with the exception that the wheel spindle nut on the right-hand side of the machine must be removed before withdrawing the complete spindle using a tommy bar inserted through the left-hand boss. There are no chain adjusters at the rear wheel frame locating holes on these models; adjustment being effected at the swinging arm pivot point.

6   Models manufactured from 1966 to 1970 featured a quickly detachable rear wheel enabling the rear wheel to be removed without disturbing the brake, sprocket or chain. To remove the wheel, simply disconnect the speedometer drive, unscrew the spindle and withdraw it from the right-hand side of the machine. Take care not to misplace the spacer that fits between the speedometer drive unit and the frame when removing the wheel.

7   To remove the brake hub and sprocket on 1966 - 1970 models, first disconnect the chain as described previously and remove the the torque arm and brake adjuster nut. Unscrew the left-hand spindle nut and withdraw the brake assembly.

6.7a Remove the brake anchor strap ...

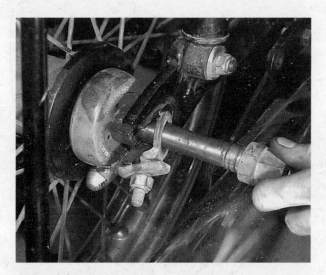

6.6a After the spindle has been withdrawn ...

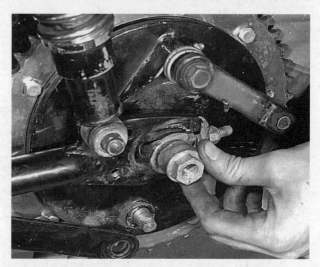

6.7b ... and left hand spindle nut ...

6.6b ... the wheel can be removed by pulling it to the right before rolling it backwards (quickly detachable type only)

6.7c ... to enable the brake hub and sprocket to be detached from the frame (1966 to 1970 models only)

7 Rear brake assembly - examination, renovation and re-assembly

1 The rear brake assembly can be withdrawn from the rear wheel, complete with brake plate, after the wheel or brake drum has been withdrawn from the frame.

2 Dismantle and inspect the brake shoes using the procedure previously described in this Chapter for single leading shoe front brakes.

7.1 Brake shoes and back plate can be withdrawn as a complete assembly from the brake drum (1966 to 1970 models)

8 Rear wheel bearings - removal, examination and replacement

1 The rear wheel bearings are of the non-adjustable ball journal type and are a drive fit within the hub. To remove, proceed as follows:-

2 On models equipped with the quickly detachable type of rear wheel (1966 to 1970) the bearing within the brake hub can be removed by first driving out the hollow spindle from the left-hand side and then removing the circlip from within the hub on the sprocket side. The bearing can then be driven out using a suitably sized piece of bar or tube.

3 To remove the bearings from the rear wheel hub, first remove the speedo drive unit and end cover and then unscrew the bearing retainer which has a **left-hand** thread and drift out the right-hand bearing by striking the hollow spindle from the left-hand side. The left-hand bearing can then be driven out from the right-hand side.

4 With machines manufactured during 1971 or later, first remove the speedometer drive unit, and spacer and unscrew the speedometer drive flange which has a **left-hand** thread. Remove the bearing retainer (right-hand thread) from inside the brake drum and drive out one of the bearings by striking the hub spacer using a drift just over ¾ inch diameter. Remove the inner abutment ring or grease seal from the side the bearing has been removed and use the spacer to drive out the remaining bearing, (take care not to damage the spacer).

5 The rear wheel bearings on the earlier 'Star' machines can be removed by removing the speedometer drive and left-hand bearing retainer and driving out the right-hand bearing by striking the spindle in that direction using a hide-faced mallet. The remaining bearing can be driven out in a similar manner.

6 The rear wheel bearings are of the journal ball type, similar to those fitted to the front wheel. Replacement is necessary if play is evident or if there is any roughness when the bearings are turned, after they have been washed out with petrol.

7 Grease the bearings as described in Section 4-3 and re-assemble hub in the reverse order of dismantling.

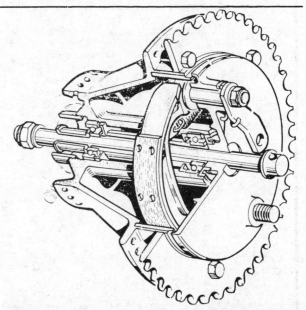

Fig. 5.6. Rear brake assembly fitted to all 1971 and later models

8.2 This hollow spindle should be driven out first before attempting to remove the circlip (1966 to 1970 models only)

8.3 The speedometer drive assembly pulls straight off, grease well before reassembly

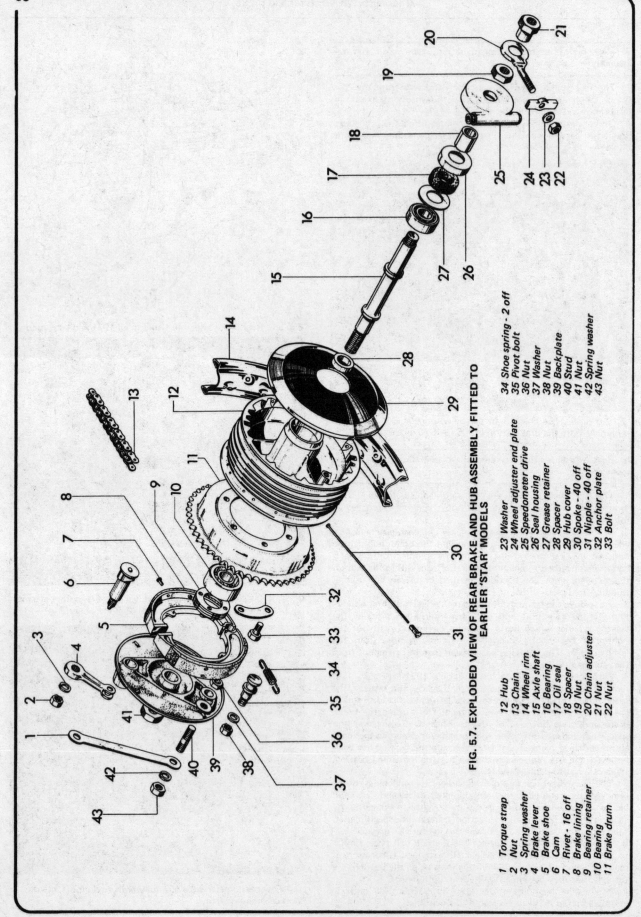

FIG. 5.7. EXPLODED VIEW OF REAR BRAKE AND HUB ASSEMBLY FITTED TO
EARLIER 'STAR' MODELS

1 Torque strap
2 Nut
3 Spring washer
4 Brake lever
5 Brake shoe
6 Cam
7 Rivet - 16 off
8 Brake lining
9 Bearing retainer
10 Bearing
11 Brake drum

12 Hub
13 Chain
14 Wheel rim
15 Axle shaft
16 Bearing
17 Oil seal
18 Spacer
19 Nut
20 Chain adjuster
21 Nut
22 Nut

23 Washer
24 Wheel adjuster end plate
25 Speedometer drive
26 Seal housing
27 Grease retainer
28 Spacer
29 Hub cover
30 Spoke - 40 off
31 Nipple - 40 off
32 Anchor plate
33 Bolt

34 Shoe spring - 2 off
35 Pivot bolt
36 Nut
37 Washer
38 Nut
39 Backplate
40 Stud
41 Nut
42 Spring washer
43 Nut

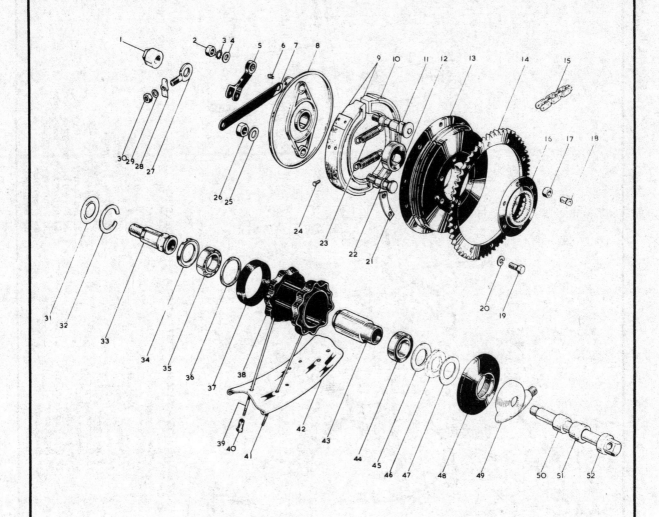

**FIG. 5.8. EXPLODED VIEW OF REAR HUB AND BRAKE ASSEMBLY FITTED TO
1966 TO 1970 MODELS**

1 Nut	14 Chainwheel (47T)	27 Adjuster	40 Nipple
2 Nut	15 Chain	28 Adjuster cap	41 Spoke (RH)
3 Lockwasher	16 Driving flange	29 Washer	42 Rim
4 Washer	17 Nut	30 Nut	43 Sleeve
5 Brake lever	18 Bolt	31 Shim	44 Bearing
6 Grease nipple	19 Bolt	32 Retainer	45 Retainer
7 Anchor strap	20 Spring washer	33 Fixed spindle	46 Felt washer
8 Cover plate	21 Lock plate	34 Retainer	47 Retainer
9 Brake shoe	22 Fulcrum pin	35 Bearing	48 End cover
10 Brake lining	23 Spring	36 Thrust washer	49 Speedometer gearbox
11 Brake cam	24 Rivet	37 Rubber seal	50 Spacer (inner, RH)
12 Bearing	25 Washer	38 Hub complete	51 Spacer (outer, RH)
13 Brake drum	26 Nut	39 Spoke (LH)	52 Spindle

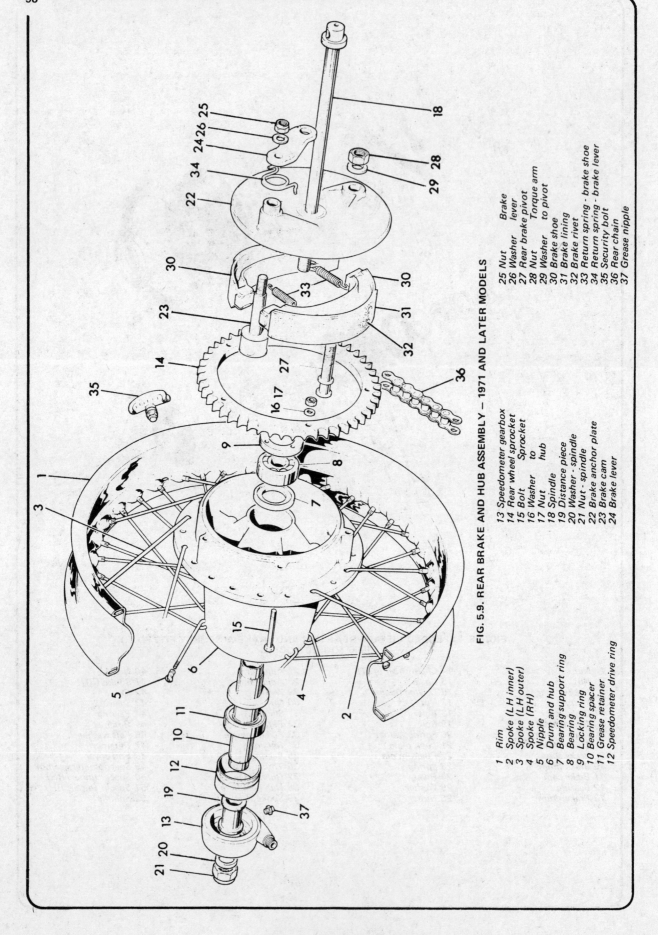

FIG. 5.9. REAR BRAKE AND HUB ASSEMBLY – 1971 AND LATER MODELS

1 Rim
2 Spoke (LH inner)
3 Spoke (LH outer)
4 Spoke (RH)
5 Nipple
6 Drum and hub
7 Bearing support ring
8 Bearing
9 Locking ring
10 Bearing spacer
11 Grease retainer
12 Speedometer drive ring

13 Speedometer gearbox
14 Rear wheel sprocket
15 Bolt
16 Washer   to   Sprocket
17 Nut   hub
18 Spindle
19 Distance piece
20 Washer - spindle
21 Nut - spindle
22 Brake anchor plate
23 Brake cam
24 Brake lever

25 Nut   Brake
26 Washer   lever
27 Rear brake pivot
28 Nut   Torque arm
29 Washer   to pivot
30 Brake shoe
31 Brake lining
32 Brake rivet
33 Return spring - brake shoe
34 Return spring - brake lever
35 Security bolt
36 Rear chain
37 Grease nipple

9 Front and rear brakes - adjustment

1 The front brake adjuster is located on the bottom fork leg on single leading shoe brakes and an additional adjuster built- in to the end of the brake operating lever on the handlebars.

2 On 1970 and earlier models fitted with twin leading shoe brakes adjustment is made at the handlebar brake lever only, the cable should be adjusted so that between 1/16 and 1/8 inch brake lever movement occurs, before the brake comes on.

No attempt should be made to alter the length of the tie rod connecting the two brake levers as this is correctly set at the factory and must not be changed.

3 1971 and later models fitted with twin leading shoe brakes have additional cam adjusters within the brake drum. To adjust, first slacken the handlebar adjuster and remove the rubber plug from the brake drum. Using a screwdriver, turn one of the cams clockwise until the brake is firmly on, then turn the cam back until the wheel is just free to rotate. Repeat this procedure on the other adjuster and then reset the cable tension by means of the handlebars adjuster.

4 The rear brake is adjusted by means of an adjusting nut on the end of the brake operating rod. Adjustment is largely a matter of personal choice, but excessive pedal travel should be avoided before the brake is fully applied .

Efficient brakes depend on good leverage of the brake operating arm. The angle between the brake operating arm and the rod should never exceed 90° when the brake is applied fully.

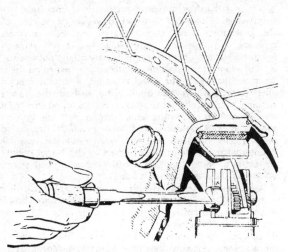

Fig. 5.10. Front shoe brake adjustment - later models only

10 Rear wheel sprocket - removal, examination and replacement

1 On the earlier 'Star' machine the sprocket and brake drum are an integral part, the drum is bolted to the wheel hub with six bolts.

2 The sprocket on machines manufactured between 1966 and 1970 can be detached from the brake drum after first removing the ten retaining bolts.

3 On 1971 and later models the sprocket is fastened to the hub by five nuts and bolts. A tubular spanner or socket will be needed to hold the nuts through the spokes on the right-hand side of the wheel. On all models the rear wheel must be taken off before the sprocket can be removed.

4 Check the condition of the sprocket teeth. If they are hooked, chipped or badly worn, the sprocket must be renewed.

5 It is bad practice to renew one sprocket on its own. The final drive sprockets should always be replaced as a pair and a new chain fitted, otherwise rapid wear will necessitate even earlier replacement next time.

11 Final drive chain - examination, lubrication and adjustment

1 The final drive chain does not have the benefit of full enclosure that is afforded to the primary drive chain. In consequence it will require attention from time to time, particularly when the machine is used on wet or dirty roads.

2 Chain adjustment is correct when there is approximately 1.1/8" play in the middle of the run, (1¾" on cam adjustment models). Always check at the tightest spot on the chain run, under load.

3 If the chain is too slack, adjustment is effected on pre-1971 models by slackening the rear wheel spindle, brake adjusting nut and torque arm, then drawing the wheel backwards by means of the chain adjusters at the end of the rear fork. Make sure each adjuster is turned an equal amount, so that the rear wheel is kept centrally-disposed within the frame. When the correct adjusting point has been reached, push the wheel hard forward to take up any slack, then tighten the spindle, not forgetting the torque arm nut. Re-check the chain tension and the wheel alignment, before the final tightening of the spindle and nuts. Re-adjust the brake if necessary.

4 On 1971 and later models the rear chain is adjusted by altering the position of the swinging arm pivot point by means of rotating cams. If the slack on the bottom run of chain exceeds about 2½ inch with the machine on its centre stand (rear suspension fully extended), the chain requires tightening.

5 Release the rear brake adjuster and separate the rod from the brake lever.

6 Loosen the right-hand swinging arm pivot nut and tap the pivot shaft through to the left side far enough to allow the cam on the left side to clear the peg.

7 Pull the swing arm to the rear and reposition the cam on both left and right sides so that a minimum of 1¾ inches of slack remains at the midpoint of the bottom run of the chain.

8 Make sure that both cams are in the same position so that correct wheel alignment is maintained, then tighten the pivot nut. Fit the brake rod into the lever and adjust the brake.

9 Application of engine oil from time to time will serve as a satisfactory form of lubrication on machines not fitted with rear chain drip feed, but it is advisable to remove the chain every 2000 miles and clean it in a bath of paraffin before immersing it in a special chain lubricant such as 'Linklyfe' or 'Chainguard'. These latter types of lubricant achieve better and more lasting penetration of the chain links and rollers are less likely to be thrown off when the chain is in motion.

10 To check whether the chain is due for replacement, lay it lengthwise in a straight line and compress it, so that all play is taken up. Anchor one end and then pull on the other, to stretch the chain in the opposite direction. If the chain extends by more than ¼" per foot, replacement is necessary.

11 When replacing the chain, make sure the spring link is positioned correctly, with the closed end facing the direction of travel. Reconnection is made easier if the ends of the chain are pressed into the rear wheel sprocket.

12 Rear wheel - replacement

1 The rear wheel is replaced in the frame by reversing the dismantling procedure described in Section 6.

2 On models fitted with the quickly detachable rear wheel make sure the distance piece is fitted between the brake plate and the right-hand side of the frame before the wheel spindle is inserted. Check that the brake torque arm has been replaced and that all bolts and nuts are tightened fully.

3 If the rear wheel sprocket has been removed either with or without the rear wheel attached, make sure that the nuts retaining the sprocket to the hub are tightened fully. If these nuts work loose, they will place a shear stress on the retaining bolts leading to their early failure.

11.11a Push chain well down into the sprocket before inserting the joining link

11.11b Closed end of spring link should always face the direction of travel

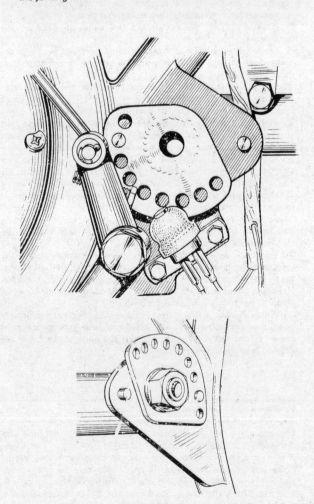

Fig. 5.11. Cam type rear chain adjuster - 1971 and later models only

13 Wheel balance

1   On any high performance machine it is important that the front wheel is balanced, to offset the weight of the tyre valve. If this precaution is not observed, the out-of-balance wheel will produce an unpleasant hammering that is felt through the handlebars at speeds from approximately 50 mph upwards.

2   To balance the front wheel, place the machine on the cnetre stand so that the front wheel is well clear of the ground and check that it will revolve quite freely, without the brake shoes rubbing. In the unbalanced state, it will be found that the wheel always come to rest in the same position, with the tyre valve in the six o'clock position. Add balance weights to the spokes diametrically opposite the tyre valve until the tyre valve is counterbalanced, then spin the wheel to check that it will come to rest in a random position on each occasion. Add or subtract weight until perfect balance is achieved.

3   Only the front wheel requires attention. There is little point in balancing the rear wheel (unless both wheels are completely interchangeable) because it will have little noticeable effect on the handling of the machine.

4   Balance weights of various sizes that will fit around the spoke nipples were orignally available from BSA Motor Cycles. If difficulty is experienced in obtaining them, lead wire or even strip solder can be used as an alternative, kept in place with insulating tape.

14 Tyres - removal and replacement

1   At some time or other the need will arise to remove and replace the tyres, either as the result of a puncture or because replacements are necessary to offset wear. To the inexperienced, tyre changing represents a formidable task yet if a few simple rules are observed and the technique learned, the whole operation is surprisingly simple.

2   To remove the tyre from either wheel, first detach the wheel from the machine by following the procedure in Chapters 5.2 or 6.6 depending on whether the front of the rear wheel is involved. Deflate the tyre by removing the valve insert and when it is fully deflated, push the bead from the tyre away from the wheel rim on both sides so that the bead enters the centre well of the rim. Remove the locking cap and push the tyre valve into the tyre itself.

3   Insert a tyre lever close to the valve and lever the edge of the tyre over the outside of the wheel rim. Very little force should

be necessary; if resistance is encountered it is probably due to the fact that the tyre beads have not entered the well of the wheel rim all the way round the tyre.

4 Once the tyre has been edged over the wheel rim, it is easy to work around the wheel rim so that the tyre is completely free on one side. At this stage, the inner tube can be removed.

5 Working from the other side of the wheel, ease the other edge of the tyre over the outside of the wheel rim that is furthest away. Continue to work around the rim until the tyre is free completely from the rim.

6 If a puncture has necessitated the removal of the tyre, re-inflate the inner tube and immerse it in a bowl of water to trace the source of the leak. Mark its position and deflate the tube. Dry the tube and clean the area around the puncture with a petrol soaked rag. When the surface has dried, apply rubber solution and allow this to dry before removing the backing from the patch and applying the patch to the surface.

7 It is best to use a patch of the self-vulcanising type, which will form a very permanent repair. Note that it may be necessary to remove a protective covering from the top surface of the patch, after it has sealed in position. Inner tubes made from synthetic rubber may require a special type of patch and adhesive, if a satisfactory bond is to be achieved.

8 Before replacing the tyre, check the inside to make sure the agent that caused the puncture is not trapped. Check the outside of the tyre, particularly the tread area, to make sure nothing is trapped that may cause a further puncture.

9 If the inner tube has been patched on a number of past occasions, or if there is a tear or large hole, it is preferable to discard it and fit a replacement. Sudden deflation may cause an accident particularly if it occurs with the front wheel.

10 To replace the tyre, inflate the inner tube sufficiently for it to assume a circular shape but only just, then push it into the tyre so that it is enclosed completly. Lay the tyre on the wheel at an angle and insert the valve through the rim tape and the hole in the wheel rim. Attach the locking cap on the first few threads, sufficient to hold the valve captive in its correct location.

11 Starting at the point furthest from the valve, push the tyre bead over the edge of the wheel rim until it is located in the central well. Continue to work around the tyre in this fashion until the whole of one side of the tyre is on the rim. It may be necessary to use a tyre lever during the final stages.

12 Make sure there is no pull on the tyre valve and again commencing with the area furthest from the valve, ease the other bead of the tyre over the edge of the rim. Finish with the area close to the valve, pushing the valve up into the tyre until the locking cap touches the rim. This will ensure the inner tube is not trapped when the last section of the bead is edged over the

rim with a tyre lever.

13 Check that the inner tube is not trapped at any point. Re-inflate the inner tube, and check the tyre is seating correctly around the wheel rim. There should be a thin rib moulded around the wall of the tyre on both sides, which should be equidistant from the wheel rim at all points. If the tyre is unevenly located on the rim, try bouncing the wheel when the tyre is at the recommended pressure. It is probable that one of the beads has not pulled clear of the centre well.

14 Always run the tyres at the recommended pressures and never under or over-inflate. The correct pressures for solo use are 22 psi front and 24 psi rear. If a pillion passenger is carried increase the rear tyre pressure to 28 psi.

15 Tyre replacement is aided by dusting the side walls, particularly in the vicinity of the beads, with a liberal coating of french chalk. Washing-up liquid can also be used to good effect, but this has the disadvantage of causing the inner surfaces of the wheel rim to rust.

16 Never replace the inner tube and tyre without the rim tape in position. If this precaution is overlooked there is a good chance of the ends of the spoke nipples chafing the inner tube and causing a crop of punctures.

17 Never fit a tyre that has a damaged ttead or sidewalls. Apart from the legal aspects there is a very great risk of a blow—out, which can have a serious consequences on any two-wheel vehicle.

18 Tyre valves rarely give trouble, but it is always advisable to check whether the valve itself is 'leaking before removing the tyre. Do not forget to fit the dust cap, which forms an effective second seal.

15 Tyre valve dust caps

1 Tyre valve dust caps are often left off when a tyre has been replaced, despite the fact that they serve an important two-fold function. Firstly they prevent dirt or other foreign matter from entering the valve and causing the valve to stick open when the tyre pump is next applied. Secondly, they form an effective second seal so that in the event of the tyre valve sticking, air will not be lost.

2 Isolated cases of sudden deflation at high speeds have been traced to the omission of the dust cap. Centrifugal force has tended to lift the tyre valve off its seating and because the dust cap is missing, there has been no second seal. Racing inner tubes contain provision for this happening because the valve inserts are fitted with stronger springs, but standard inner tubes do not, hence the need of the dust cap.

3 Note that when a dust cap is fitted for the first time, the wheel may have to be rebalanced.

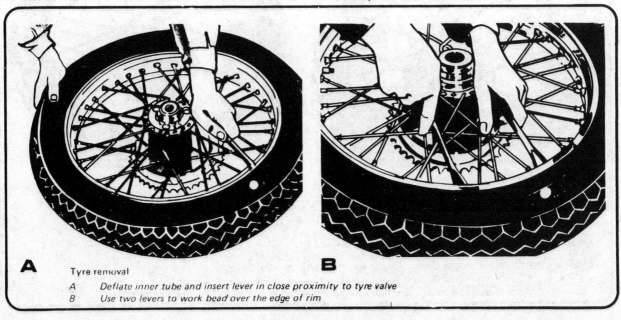

A   Tyre removal
A   Deflate inner tube and insert lever in close proximity to tyre valve
B   Use two levers to work bead over the edge of rim

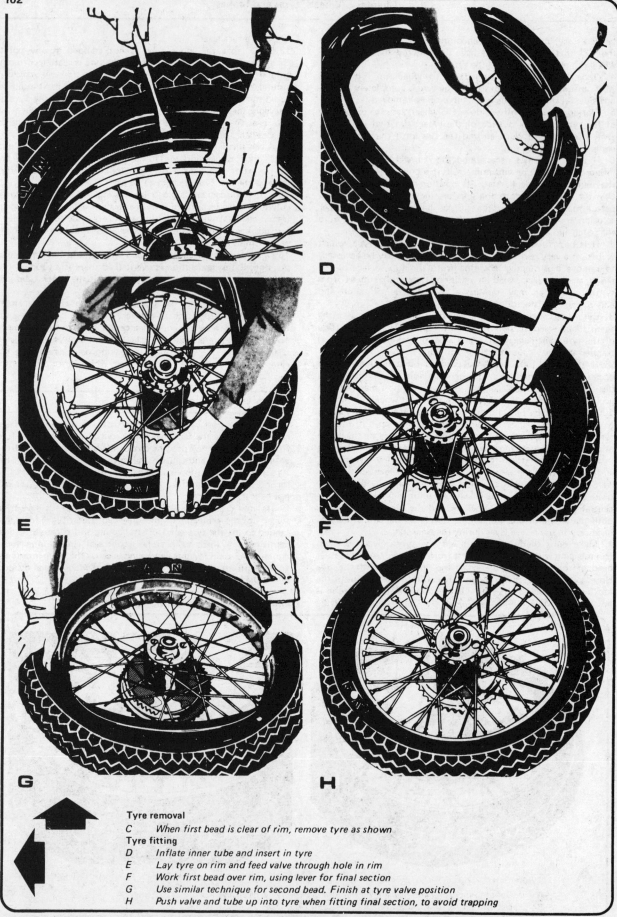

Tyre removal

C   *When first bead is clear of rim, remove tyre as shown*

Tyre fitting

D   *Inflate inner tube and insert in tyre*

E   *Lay tyre on rim and feed valve through hole in rim*

F   *Work first bead over rim, using lever for final section*

G   *Use similar technique for second bead. Finish at tyre valve position*

H   *Push valve and tube up into tyre when fitting final section, to avoid trapping*

16 Fault diagnosis - Wheels, brakes and tyres

Symptom	Reason/s	Remedy
Handlebars oscillate at low speeds	Buckle or flat in wheel rim	Check rim alignment by spinning. Correct by retensioning spokes or by having wheel rebuilt on new rim
	Tyre not straight on rim	Check tyre alignment
Machine lacks power and accelerates poorly	Brakes binding	Warm brake drums provide best evidence. Re-adjust brakes
Brakes grab, even when applied gently	Ends of brake shoes not chamfered	Chamfer with file
	Elliptical brake drum	Lightly skim in lathe (specialist attention required)
Brake pull-off sluggish	Brake cam binding in housing	Free and grease
	Weak brake shoe springs	Replace, if springs not displaced
Harsh transmission	Worn or badly adjusted chains	Replace or adjust, as necessary
	Hooked or badly worn sprockets	Replace as a pair
	Rear wheel sprocket nuts loose	Check and tighten
Middle of tyre treads wear rapidly	Tyres over-inflated	Check and readjust pressures
Edges of tyre treads wear rapidly	Tyres under inflated	Check and increase pressures
Forks hammer at high speeds	Front wheel not balanced	Balance wheel by adding balance weights

Chapter 6 Electrical equipment

Contents

Specifications

Alternator

Make ... ...   Lucas (all models)

Battery (lead acid)

Make ... ...   Lucas ML9E 6 volt - 1958 - 1966 'Star' models
    Lucas PUZ5A 12 volt - all 1967 and later models

Voltage ... ...   6 volt - all 1958 - 1966 models
    12 volt - 1967 onwards

Earth connection ... ...   Positive earth - all models

Lighting

Headlamp bulb ... ...   30/24W pre-focus - 1958 - 1966 'Star' models (6 volt)
    50/40 pre-focus - 1967 and later models (12 volt)

Pilot lamp bulb ... ...   6/18W 1958 - 1966 'Star' models (6 volt)
Rear/stop lamp bulb ... ...   6/21W 1967 and later models (12 volt)

Speedometer bulb ... ...   2.2W - 1958 - 1966 'Star' models (6 volt)
    6W - 1967 and later models (12 volt)

Fuse

Loading ... ...   35 amps - all models except the earlier 'Star' series
Location ... ...   Negative lead from battery

1 General description

1 All BSA single cylinder unit construction machines are fitted with a crankshaft driven alternator for the purpose of supplying electrical current to charge the battery and also to provide the 'kick-over' current for emergency starting purposes, i.e. flat battery. The alternator is very simple in construction and comprises only two basic parts; the stator, (which as its name implies does not move) consists of induction coils encased in a laminated ring structure and bolted onto the front of the crankcase. The rotor, comprising several permanent magnets within an aluminium casting, is keyed to the crankshaft and rotates within the stator when the engine is running, inducing an alternating current, (AC).

2 As there are no brushes, bearing or commutator to wear out the alternator requires no maintenance, the only check necessary is to ensure that the rotor circumferance does not touch the stator at any point. If the rotor is removed it is not necessary to place 'keepers' across the magnet poles. The AC output from the alternator is converted to DC by a rectifier. On earlier 6 volt models the alternator output varies with battery demand: if the headlights are switched on maximum current is produced where as during the daylight running the alternator provides just enough current to charge the battery. On the earlier 'Star' models the high output can be obtained by selecting the ignition switch to 'emergency' to supply sufficient current to the coil for engine starting purposes, (see Chapter 3).

Models produced during 1966 and after are fitted with a 12 volt electrical system, the alternator works at continuous maximum output and the voltage to the battery is regulated by a Zener diode. Because the low speed output of the 12 volt alternator is constant the machine can be started with a flat battery without the necessity of an emergency start switch. Some later models are fitted with a capacitor which can be connected up in place of the battery for field or track events for weight saving purposes.

2 Alternator - checking output

1 The electrical output from the alternator fitted to the BSA single models can be checked only with specialised equipment of the multi-meter type. If the alternator is suspect, it should be checked by either a BSA agent or an auto-electrical expert.

3 Battery - examination and maintenance

1 Batteries of different amp/hour capacities have been fitted to the various BSA single models, the capacity of the battery being dependant on the current rating of the full electrical load. A battery should always be replaced with another of similar capacity, it is a false economy to fit a battery having a lower capacity because the cost advantage will soon be offset by the reduced working life. Obviously the new battery must be the same voltage, i.e. 6 or 12 volt depending on year of manufacture.
2 Maintenance is normally limited to keeping the electrolyte level just above the plates and separators. Modern batteries have translucent plastic cases, which make the check of electrolyte level much easier.
3 Unless acid is spilt, which may occur if the machine falls over, the electrolyte should always be topped up with distilled water until the correct level is restored. If acid is spilt on any part of the machine, it should be neutralised with an alkali such as washing soda and washed away with plenty of water, otherwise serious corrosion will occur. Top up with sulphuric acid of the correct specific gravity (1.260 - 1.280) ONLY when spillage has occurred.
4 It is seldom practicable to repair a cracked battery case because the acid already in the joint will prevent the formation of an effective seal. It is always best to replace a cracked battery, especially in view of the corrosion that will be caused by the leakage of acid.
5 The Varley battery, fitted to some models is of the lead/acid type, but has the electrolyte absorbed in glass wool packing that surrounds the plates. This obviates the risk of acid spillage when the battery is inverted. Topping up consists of adding a teaspoonful fo distilled water to each cell and leaving the battery to stand for approximately five minutes before draining off any excess.
6 Always ensure that the teminals are clean, tight and free from corrosion. After cleaning the terminals, smear them with petroleum jelly to prevent further corrosion.

3.2 Check the level of the battery regularly, do NOT overfill

4 Battery - charging procedure

1 Whilst the machine is running, it is unlikely that the battery will require attention other than routine maintenance because the alternator will keep it charged. However, if the machine is used for a sucession of short journeys only, mainly during the hours of darkness when the lights are in full use, it is possible that the output from the generator will be able to keep pace with the heavy electrical demand, especially on the earlier models. Under these circumstances it will be necessary to remove the battery from time to time, to have it charged independantly.
2 The normal charging rate is 1 amp. A more rapid charge can be given in an emergency, but this should be avoided if possible because it will shorten the working life of the battery.
3 When the battery has been removed from a machine that has been laid up, a 'refresher' charge should be given every six weeks if the battery is to be maintained in good condition.

5 Selenium rectifier - general description

1 The function of the selenium rectifier is to convert the AC produced by the alternator to DC so that it can be used to charge the battery and operate the lighting circuit etc. The usual sympton of a defective rectifier is a battery that discharges rapidly because it is receiving no charge from the generator.
2 The rectifier is located where it is not exposed to water or battery acid, which will cause it to malfunction. The question of access is of relatively little importance because the rectifier is unlikely to give trouble during normal operating conditions It is not practicable to repair a damage rectifier; replacement is the only satisfactory solution. One of the most frequent causes of rectifier failure is the inadvertant connection of the battery in reverse, which results in a reverse current flow.
3 It is not possible to check whether the rectifier is functioning correctly without the appropriate test equipment. A BSA agent or an auto-electrical expert are best qualified to advise in such cases.
4 Do not loosen the rectifier locking nut or bend, cut, scratch or rotate the selenium wafers. Any such action will cause the electrode alloy coating to peel and destroy the working action.

6 Zener diode - general description

1 The Zener diode is a compact and extremely efficient version of the bulky twin coil voltage regulators fitted to much earlier machines.
2 If the battery voltage is low, the diode's corresponding leakage resistance to earth via the heat sink is high and the full output from the alternator passes to the battery. The battery, (providing it is in good condition) soon reaches a full state of charge and battery voltage across the diode begins to rise until at 13.5 volts the resistance value of the diode drops sufficiently to allow some of the charging current to by-pass to the heat sink, (earth). As the voltage continues to rise the conductivity of the diode increases rapidly until at approximately 15.5 volts about 5 amps of the alternator output is being bypassed through the heat sink, thus ensuring that the battery is not overcharged.
3 When the battery voltage is reduced due to the use of the headlamp or other electrical equipment the resistance to earth of the diode increases, allowing the full output of the alternator to replenish the demands made on the battery.
4 On 1970 and earlier models the Zener diode and its associated heat sink are located below the steering head and is retained by a single nut and bolt. On 1971 and later models the diode is contained in the electrical box beneath the front of the petrol tank and utilises the lid of the box as a heat sink.
5 No maintenance of the diode is necessary apart from ensuring the finned heat sink on earlier models is kept clean and on later models the diode is in good contact with the electrical box lid. If it is suspected that the diode is at fault, it should be taken to an auto-electrical dealer for checking and if necessary replaced.

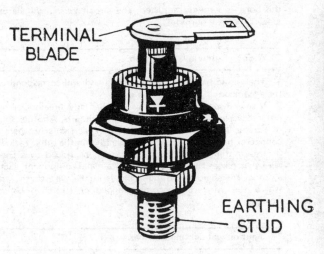

TERMINAL BLADE

EARTHING STUD

Fig. 6.1. The Zener diode fitted to all 12 volt models

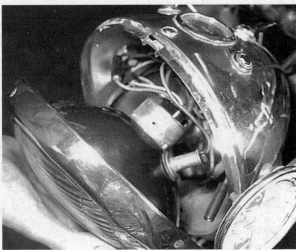

7.1 Headlamp is released after slackening the top clamping screw

7 Headlamp - replacing bulbs and adjusting beam height

1 To remove the headlamp rim, unscrew the screw at the top of the headlamp shell. The rim will then pull off, complete with the reflector unit and bulbs.

2 The reflector unit contains a double-filament bulb that provides the main and dipped headlamp beams, and a pilot lamp bulb for parking use. The double-filament headlamp bulb is controlled from a dip switch mounted on the handlebars. The pilot lamp is found adjacent to the main bulb in a holder which is a push fit in the reflector. On some models, oil pressure, main beam and direction indicator warning lights are fitted on the inside of the headlamp shell. The bulb holders are a push fit and the bulbs are the bayonet type.

3 It is not necessary to refocus the headlamp when a new bulb is fitted because the bulbs are of the prefocus type. To release the bulb holder from the reflector, twist and pull.

4 Beam height is adjusted by slackening the two headlamp shell retaining bolts and tilting the headlamp either upwards or downwards. Adjustment should always be made with the rider normally seated.

5 UK lighting regulations stipulate that the lighting system must be arranged so that the light will not dazzle a person standing in the same horizontal plane as the vehicle at a distance greater than 25 yards from the lamp, whose eye level is not less than 3 feet 6 inches above that plane. It is easy to approximate this setting by placing the machine 25 yards away from a wall, on a level road, and setting the beam height so that it is concentrated at the same height as the distance from the centre of the headlamp to the ground. The rider must be seated normally during this operation and also the pillion passenger, if one is carried regularly.

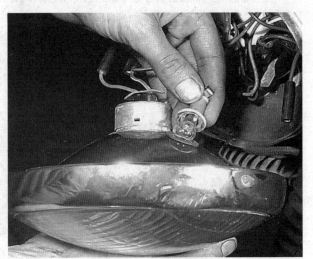

7.2 Pilot bulb is a push fit in the reflector

8 Tail and stop lamp - replacing bulbs

1 The early 'Star' models have a double filament bulb of 6 volt 6/18W rating and later models use a double filament 12 volt 6/21W bulb to illuminate the rear of the machine and the rear number plate, and to give visual warning when the rear brake is applied. To gain access to the bulb, unscrew the two screws that retain the moulded plastics lens cover to the rear lamp assembly and remove the cover and sealing gasket.

2 The stop lamp is illuminated by a micro-switch on the left-hand side of the machine, operated by the brake rod via a lever. The switch does not require any special attention apart from cleaning off any accumulated road dirt.

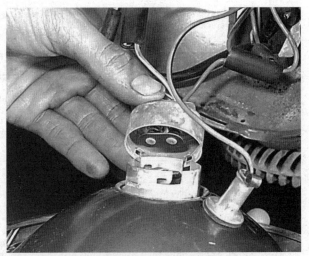

7.3a The main bulb retainer has three offset slots to ensure correct position of contacts ...

7.3b ... and the pre-focus bulb has a small cut-out that aligns with a dimple in the reflector shroud

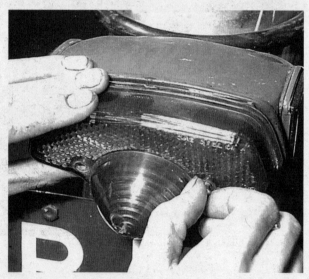

8.1 Do not overtighten the two rear lamp screws - the plastic lens is easily cracked

9 Speedometer and tachometer bulb - replacement

1 The bulb that illuminates the dial of the speedometer and, if fitted the tachometer has a bayonet fitting in a metal bulb-holder that bushes into the base of each instrument case. The bulbs are rated at 2.2W each for 6 volt models and 6W for 12 volt machines.

10 Horn - adjustment

1 A horn of the electro-magnetic type is fitted to every machine usually below the tank or dual seat. It is operated from a push button, mounted on the handlebars.
2 A small serrated screw located near the terminals affords a means of adjustment. If the horn does not function, turn this screw anti-clockwise until the restored note just fails to sound and then turn it back clockwise about one quarter of a turn.
3 When testing the horn during adjustment, do not operate the horn button any more than absolutely necessary otherwise, if the horn is wrongly adjusted, the circuit wiring will be overloaded and may burn out.

11 Wiring - layout and inspection

1 The wiring harness is colour-coded and will correspond with the accompanying wiring diagrams.
2 Visual inspection will show whether any breaks or frayed outer coverings are giving rise to short circuits. Another source of trouble may be the snap connectors, particularly where the connector has not been pushed home fully in the outer casing.
3 Intermittent short circuits can often be traced to a chafed wire that passes through or close to a metal component, such as a frame member. Avoid tight bends in the wire or situations where the wire can become trapped or stretched between casings.

12 Ignition and lighting switches

1 On the earlier 'Star' models the ignition switch is on the side panel behind the engine and, as described in Chapter 3, has two positions one for normal running and the other for emergency starting. Very early machines were not fitted with a key; the complete knob being rotated to the required position. All later models have a detachable ignition key, 1966 and later models are fitted with high output 12 volt alternator and an emergency start position is unnecessary. The 'star' models have a rotary lighting switch on the headlamp shell for selecting side or headlight while later machines are fitted with a toggle switch. The dipswitch is on the handlebars adjacent to the horn button. A full beam warning light is provided on the headlamp shell.
2 1971 and later models are fitted with two multi-purpose switches on the handlebars. The left-hand switch incorporates a headlamp dipping lever switch, a headlamp flasher pushbutton and a horn pushbutton. The right-hand switch houses the flashing indicators control switch and the ignition cut-out button. The switches are quite complex in construction and it is not advisable to attemtp to dismantle them as special equipment is required for reassembly. If trouble is experienced with any of the switch circuits, and auto-electrical firm or BSA dealer should be consulted and the complete switch assembly must be replaced if necessary.
3 On no account oil the switch or oil will spread across the internal contacts, to form an effective insulator.

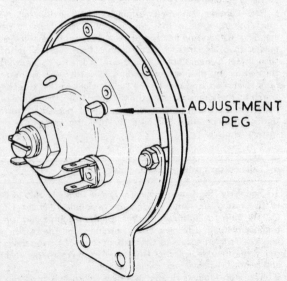

ADJUSTMENT PEG

Fig. 6.2. Electric horn adjuster

13 FAULT DIAGNOSIS - Electrical equipment

Symptom	Reason/s	Remedy
Tail or stop lamp does not illuminate	Burnt out bulb fitment or dirty contacts	Clean spring-loaded contacts and replace bulb if necessary.
Headlamp illuminated on either dipped or main beam only	As above	As above
Horn note is barely audible or non-existant	Faulty connection or out of adjustment	Check cleanliness and security of terminals and, if necessary adjust horn as described in Section 10.
Lights dim when engine revs. drop	Corroded battery terminals or low electrolyte level	Clean and tighten battery terminals and check acid level. If fault persists take battery to a garage and have it tested. If faulty replace it.
Battery is known to be in good condition but fails to hold a charge	Possibly alternator faulty	Take machine to an auto-electrical dealer and have the alternator output checked. If satisfactory have the rectifier and Zener diode checked and replace if necessary.

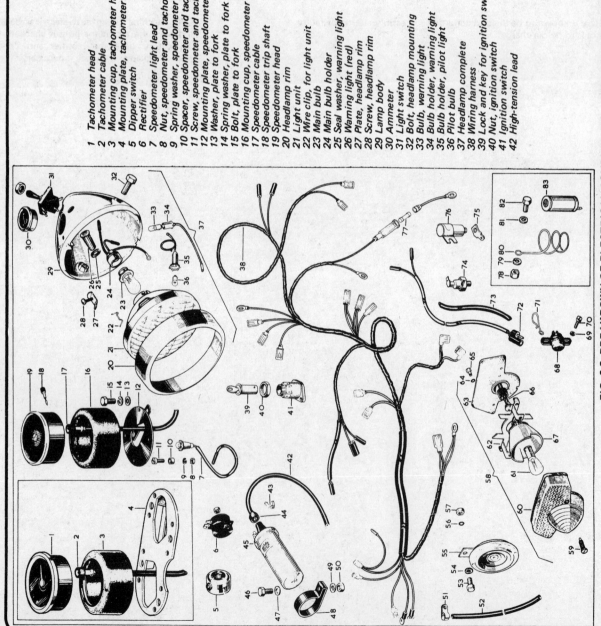

1 Tachometer head
2 Tachometer cable
3 Mounting cup, tachometer head
4 Mounting plate, tachometer and speedometer
5 Dipper switch
6 Rectifier
7 Speedometer light lead
8 Nut, speedometer and tachometer
9 Spring washer, speedometer and tachometer
10 Spacer, speedometer and tachometer
11 Screw, speedometer and tachometer
12 Mounting plate, speedometer only
13 Washer, plate to fork
14 Spring washer, plate to fork
15 Bolt, plate to fork
16 Mounting cup, speedometer head
17 Speedometer cable
18 Speedometer trip shaft
19 Speedometer head
20 Headlamp rim
21 Light unit
22 Wire clip, for light unit
23 Main bulb
24 Main bulb holder
25 Seal washer, warning light
26 Warning light (red)
27 Plate, headlamp rim
28 Screw, headlamp rim
29 Lamp body
30 Ammeter
31 Light switch
32 Bolt, headlamp mounting
33 Bulb, warning light
34 Bulb holder, warning light
35 Bulb holder, pilot light
36 Pilot bulb
37 Headlamp complete
38 Wiring harness
39 Lock and key for ignition switch
40 Nut, ignition switch
41 Ignition switch
42 High-tension lead

43 Clip for high-tension lead
44 Grommet for high tension lead
45 Ignition coil
46 Bolt, coil clip
47 Washer, coil clip
48 Coil clip
49 Spacer, coil slip
50 Nut, coil clip
51 Clip, battery vent pipe
52 Battery vent pipe
53 Bolt, horn to bracket
54 Washer, horn to bracket
55 Horn (clearhooter)
56 Lockwasher, horn to bracket
57 Nut, horn to bracket
58 Stop/tail lamp complete
59 Screw, lens fixing
60 Lens, tail lamp
61 Bulb, tail lamp
62 Base, tail lamp
63 Gasket, tail lamp
64 Lockwasher, tail lamp
65 Screw, tail lamp
66 Interior bulb holder, tail lamp
67 Bulb holder assembly
68 Stop switch, tail lamp
69 Washer for switch screw
70 Screw, stop lamp switch
71 Cable clip, stop switch leads
72 Tail lamp lead
73 Tail lamp lead tube
74 Zener diode
75 'Lucar' blade, condenser
76 Condenser, contact breaker
77 Fuse (12 volt, 35 amp)
78 Nut, capacitor mounting
79 Spring washer, capacitor mounting
80 Spring, capacitor mounting
81 Washer, capacitor mounting
82 Bolt, capacitor mounting
83 Capacitor

FIG. 6.3. BREAKDOWN OF ELECTRICAL COMPONENTS — MODELS FROM 1966 TO 1970

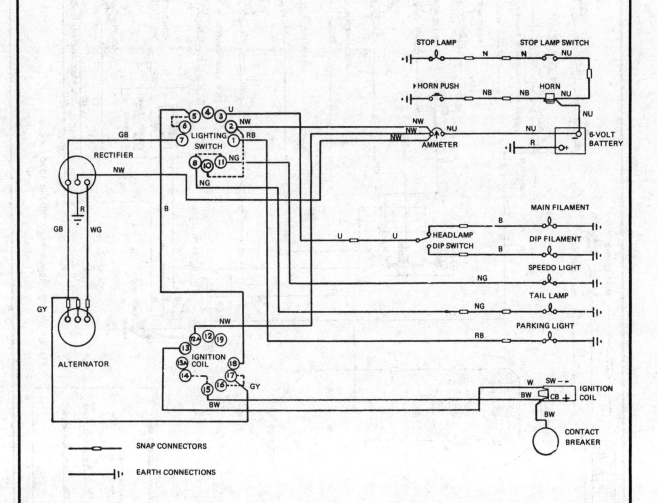

Fig. 6.4. Wiring diagram for 1958 to 1966 'Star' models

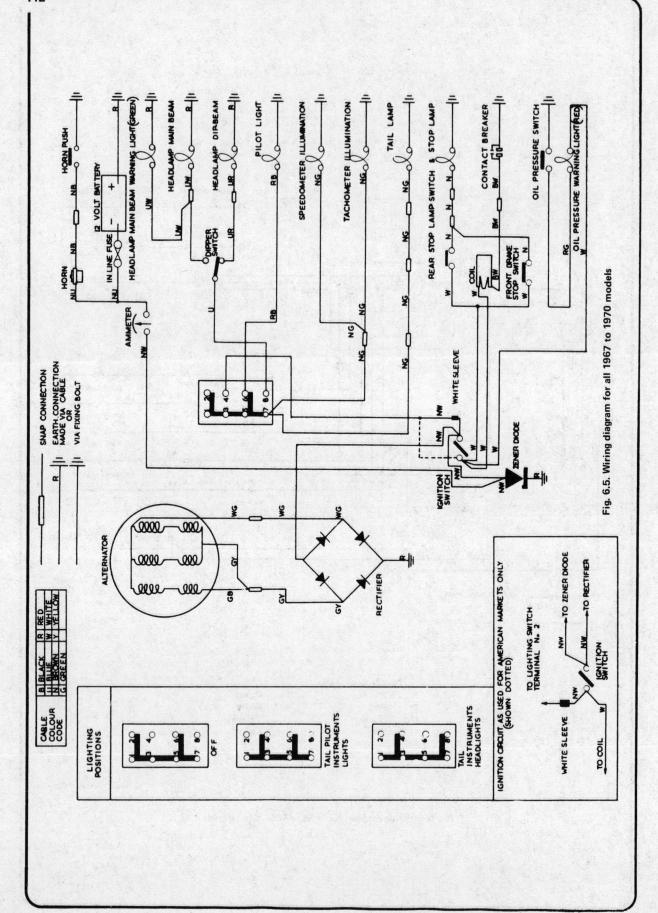

Fig. 6.5. Wiring diagram for all 1967 to 1970 models

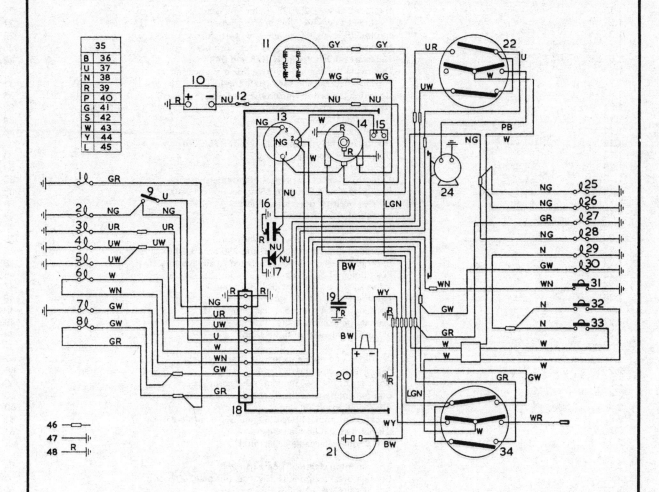

FIG. 6.6. WIRING DIAGRAM FOR ALL 1971 AND LATER MODELS

1   Left direction indicator
2   Parking light
3   Dipped headlight beam
4   Main headlight beam
5   Main beam warning light
6   Oil pressure warning light (250 cm³ only)
7   Right direction indicator
8   Direction indicator warning light
9   Headlight switch
10 Battery
11 Alternator
12 Fuse
13 Ignition/lighting switch
14 Rectifier
15 Direction indicator unit
16 Capacitor (2.M.C.)

17 Zener diode
18 Plug and socket (9 pin)
19 Ignition capacitor
20 Ignition coil
21 Contact breaker
22 Left handlebar switches
24 Horn
25 Speedometer light
26 Tachometer light
27 Left direction indicator
28 Rear light
29 Stop light
30 Right direction indicator
31 Oil pressure switch (250 cm³ only)
32 Front and rear brake light switches
33 Front and rear brake light switches

34 Right handlebar switches
35 Cable colour code
36 Black
37 Blue
38 Brown
39 Red
40 Purple
41 Green
42 Slate
43 White
44 Yellow
45 Light
46 Snap connectors
47 Ground (earth) connection via cable
48 Ground (earth) connection via fixing bolt

List of illustrations

Conversion factors

Length (distance)

Inches (in)	x 25.4	= Millimetres (mm)	x 0.0394	=	Inches (in)
Feet (ft)	x 0.305	= Metres (m)	x 3.281	=	Feet (ft)
Miles	x 1.609	= Kilometres (km)	x 0.621	=	Miles

Volume (capacity)

Cubic inches (cu in; in³)	x 16.387	= Cubic centimetres (cc; cm³)	x 0.061	=	Cubic inches (cu in; in³)
Imperial pints (Imp pt)	x 0.568	= Litres (l)	x 1.76	=	Imperial pints (Imp pt)
Imperial quarts (Imp qt)	x 1.137	= Litres (l)	x 0.88	=	Imperial quarts (Imp qt)
Imperial quarts (Imp qt)	x 1.201	= US quarts (US qt)	x 0.833	=	Imperial quarts (Imp qt)
US quarts (US qt)	x 0.946	= Litres (l)	x 1.057	=	US quarts (US qt)
Imperial gallons (Imp gal)	x 4.546	= Litres (l)	x 0.22	=	Imperial gallons (Imp gal)
Imperial gallons (Imp gal)	x 1.201	= US gallons (US gal)	x 0.833	=	Imperial gallons (Imp gal)
US gallons (US gal)	x 3.785	= Litres (l)	x 0.264	=	US gallons (US gal)

Mass (weight)

Ounces (oz)	x 28.35	= Grams (g)	x 0.035	=	Ounces (oz)
Pounds (lb)	x 0.454	= Kilograms (kg)	x 2.205	=	Pounds (lb)

Force

Ounces-force (ozf; oz)	x 0.278	= Newtons (N)	x 3.6	=	Ounces-force (ozf; oz)
Pounds-force (lbf; lb)	x 4.448	= Newtons (N)	x 0.225	=	Pounds-force (lbf; lb)
Newtons (N)	x 0.1	= Kilograms-force (kgf; kg)	x 9.81	=	Newtons (N)

Pressure

Pounds-force per square inch (psi; lbf/in²; lb/in²)	x 0.070	= Kilograms-force per square centimetre (kgf/cm²; kg/cm²)	x 14.223	=	Pounds-force per square inch (psi; lbf/in²; lb/in²)
Pounds-force per square inch (psi; lbf/in²; lb/in²)	x 0.068	= Atmospheres (atm)	x 14.696	=	Pounds-force per square inch (psi; lbf/in²; lb/in²)
Pounds-force per square inch (psi; lbf/in²; lb/in²)	x 0.069	= Bars	x 14.5	=	Pounds-force per square inch (psi; lbf/in²; lb/in²)
Pounds-force per square inch (psi; lbf/in²; lb/in²)	x 6.895	= Kilopascals (kPa)	x 0.145	=	Pounds-force per square inch (psi; lbf/in²; lb/in²)
Kilopascals (kPa)	x 0.01	= Kilograms-force per square centimetre (kgf/cm²; kg/cm²)	x 98.1	=	Kilopascals (kPa)
Millibar (mbar)	x 100	= Pascals (Pa)	x 0.01	=	Millibar (mbar)
Millibar (mbar)	x 0.0145	= Pounds-force per square inch (psi; lbf/in²; lb/in²)	x 68.947	=	Millibar (mbar)
Millibar (mbar)	x 0.75	= Millimetres of mercury (mmHg)	x 1.333	=	Millibar (mbar)
Millibar (mbar)	x 0.401	= Inches of water (inH$_2$O)	x 2.491	=	Millibar (mbar)
Millimetres of mercury (mmHg)	x 0.535	= Inches of water (inH$_2$O)	x 1.868	=	Millimetres of mercury (mmHg)
Inches of water (inH$_2$O)	x 0.036	= Pounds-force per square inch (psi; lbf/in²; lb/in²)	x 27.68	=	Inches of water (inH$_2$O)

Torque (moment of force)

Pounds-force inches (lbf in; lb in)	x 1.152	= Kilograms-force centimetre (kgf cm; kg cm)	x 0.868	=	Pounds-force inches (lbf in; lb in)
Pounds-force inches (lbf in; lb in)	x 0.113	= Newton metres (Nm)	x 8.85	=	Pounds-force inches (lbf in; lb in)
Pounds-force inches (lbf in; lb in)	x 0.083	= Pounds-force feet (lbf ft; lb ft)	x 12	=	Pounds-force inches (lbf in; lb in)
Pounds-force feet (lbf ft; lb ft)	x 0.138	= Kilograms-force metres (kgf m; kg m)	x 7.233	=	Pounds-force feet (lbf ft; lb ft)
Pounds-force feet (lbf ft; lb ft)	x 1.356	= Newton metres (Nm)	x 0.738	=	Pounds-force feet (lbf ft; lb ft)
Newton metres (Nm)	x 0.102	= Kilograms-force metres (kgf m; kg m)	x 9.804	=	Newton metres (Nm)

Power

Horsepower (hp)	x 745.7	= Watts (W)	x 0.0013	=	Horsepower (hp)

Velocity (speed)

Miles per hour (miles/hr; mph)	x 1.609	= Kilometres per hour (km/hr; kph)	x 0.621	=	Miles per hour (miles/hr; mph)

Fuel consumption*

Miles per gallon, Imperial (mpg)	x 0.354	= Kilometres per litre (km/l)	x 2.825	=	Miles per gallon, Imperial (mpg)
Miles per gallon, US (mpg)	x 0.425	= Kilometres per litre (km/l)	x 2.352	=	Miles per gallon, US (mpg)

Temperature

Degrees Fahrenheit = (°C x 1.8) + 32   Degrees Celsius (Degrees Centigrade; °C) = (°F - 32) x 0.56

It is common practice to convert from miles per gallon (mpg) to litres/100 kilometres (l/100km), where mpg x l/100 km = 282

English/American terminology

Because this book has been written in England, British English component names, phrases and spellings have been used throughout. American English usage is quite often different and whereas normally no confusion should occur, a list of equivalent terminology is given below.

English	American	English	American
Air filter	Air cleaner	Number plate	License plate
Alignment (headlamp)	Aim	Output or layshaft	Countershaft
Allen screw/key	Socket screw/wrench	Panniers	Side cases
Anticlockwise	Counterclockwise	Paraffin	Kerosene
Bottom/top gear	Low/high gear	Petrol	Gasoline
Bottom/top yoke	Bottom/top triple clamp	Petrol/fuel tank	Gas tank
Bush	Bushing	Pinking	Pinging
Carburettor	Carburetor	Rear suspension unit	Rear shock absorber
Catch	Latch	Rocker cover	Valve cover
Circlip	Snap ring	Selector	Shifter
Clutch drum	Clutch housing	Self-locking pliers	Vise-grips
Dip switch	Dimmer switch	Side or parking lamp	Parking or auxiliary light
Disulphide	Disulfide	Side or prop stand	Kick stand
Dynamo	DC generator	Silencer	Muffler
Earth	Ground	Spanner	Wrench
End float	End play	Split pin	Cotter pin
Engineer's blue	Machinist's dye	Stanchion	Tube
Exhaust pipe	Header	Sulphuric	Sulfuric
Fault diagnosis	Trouble shooting	Sump	Oil pan
Float chamber	Float bowl	Swinging arm	Swingarm
Footrest	Footpeg	Tab washer	Lock washer
Fuel/petrol tap	Petcock	Top box	Trunk
Gaiter	Boot	Torch	Flashlight
Gearbox	Transmission	Two/four stroke	Two/four cycle
Gearchange	Shift	Tyre	Tire
Gudgeon pin	Wrist/piston pin	Valve collar	Valve retainer
Indicator	Turn signal	Valve collets	Valve cotters
Inlet	Intake	Vice	Vise
Input shaft or mainshaft	Mainshaft	Wheel spindle	Axle
Kickstart	Kickstarter	White spirit	Stoddard solvent
Lower leg	Slider	Windscreen	Windshield
Mudguard	Fender		

Index

Preserving Our Motoring Heritage

< The Model J Duesenberg Derham Tourster. Only eight of these magnificent cars were ever built – this is the only example to be found outside the United States of America

Almost every car you've ever loved, loathed or desired is gathered under one roof at the Haynes Motor Museum. Over 300 immaculately presented cars and motorbikes represent every aspect of our motoring heritage, from elegant reminders of bygone days, such as the superb Model J Duesenberg to curiosities like the bug-eyed BMW Isetta. There are also many old friends and flames. Perhaps you remember the 1959 Ford Popular that you did your courting in? The magnificent 'Red Collection' is a spectacle of classic sports cars including AC, Alfa Romeo, Austin Healey, Ferrari, Lamborghini, Maserati, MG, Riley, Porsche and Triumph.

A Perfect Day Out

Each and every vehicle at the Haynes Motor Museum has played its part in the history and culture of Motoring. Today, they make a wonderful spectacle and a great day out for all the family. Bring the kids, bring Mum and Dad, but above all bring your camera to capture those golden memories for ever. You will also find an impressive array of motoring memorabilia, a comfortable 70 seat video cinema and one of the most extensive transport book shops in Britain. The Pit Stop Cafe serves everything from a cup of tea to wholesome, home-made meals or, if you prefer, you can enjoy the large picnic area nestled in the beautiful rural surroundings of Somerset.

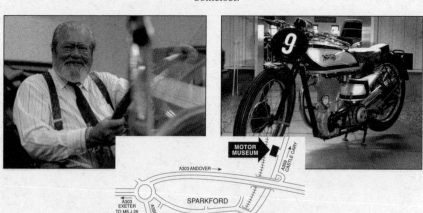

John Haynes O.B.E., Founder and Chairman of the museum at the wheel of a Haynes Light 12. >

< The 1936 490cc sohc-engined International Norton – well known for its racing success

The Museum is situated on the A359 Yeovil to Frome road at Sparkford, just off the A303 in Somerset. It is about 40 miles south of Bristol, and 25 minutes drive from the M5 intersection at Taunton.
Open 9.30am - 5.30pm (10.00am - 4.00pm Winter) 7 days a week, *except Christmas Day, Boxing Day and New Years Day*
Special rates available for schools, coach parties and outings Charitable Trust No. 292048